ε Epsilon

Fractions

Student Workbook

Math·U·See®

1-888-854-MATH (6284) - mathusee.com
sales@mathusee.com

Epsilon Student Workbook: Fractions

©2012 Math-U-See, Inc.

Published and distributed by Demme Learning.

mathusee.com

1-888-854-6284 or +1 717-283-1448 | demmelearning.com
Lancaster, Pennsylvania USA

ISBN 978-1-60826-070-6
Revision Code 1218

Printed in the United States of America by Bindery Associates LLC
  2 3 4 5 6 7 8 9 10

For information regarding CPSIA on this printed material call: 1-888-854-6284
and provide reference #1218-121018

Epsilon ε

	LESSON PRACTICE			SYSTEMATIC REVIEW			
	A	B	C	D	E	F	TEST
1 Fraction of a Number							
2 Fraction of One							
3 Add and Subtract — Equal							
4 Equivalent Fractions							
5 Add and Subtract — Unequal							
6 Rule of 4							
7 Compare Fractions							
8 Add Multiple Fractions							
Unit Test 1							
9 Multiply Fractions							
10 Dividing Fractions							
11 Common Factors							
12 Simplify Fractions 1							
13 Simplify Fractions 2							
14 Linear Measure							
15 Mixed Numbers							
16 Linear Mixed Numbers							
Unit Test 2							
17 Add Subtract Mixed Numbers							
18 Add Mixed Regroup							
19 Subtract Mixed Regroup							
20 Same Difference							
21 Add Mixed Unequal Denom.							
22 Subtract Mixed Unequal Denom.							
23 Divide with Reciprocal							
Unit Test 3							
24 Solve for Unknown 1							
25 Multiply 3 Fractions							
26 Solve for Unknown 2							
27 Circle: Area and Circumference							
28 Solve for Unknown 3							
29 Fraction, Decimal, Percent							
30 Solve for Unknown 4							
Unit Test 4							
Final Test							

APPLICATION AND ENRICHMENT PAGES

This edition of the *Epsilon Student Workbook* includes extra activity pages titled "Application and Enrichment." You will find one enrichment page after the last systematic review page for each lesson. These activities are intended to do the following:

- Provide a variety of ways to practice lesson concepts.
- Lead students through detailed strategies for solving fraction word problems.
- Stimulate thinking by presenting concepts in different formats.
- Introduce new concepts that may be useful to students at this level.
- Enrich learning with additional age-appropriate activities.

One of the main goals of these pages is to provide more help in learning how to solve fraction word problems. The Application and Enrichment pages also introduce some terms and ideas that your child may see on standardized tests. Mastery of these concepts is *not* necessary to move to the next level of Math-U-See. However, exposure to the concepts may be useful to some students.

Many of these activities will be challenging because they require a new way of looking at a concept. Encourage students to think carefully for themselves, but do not hesitate to give them as much help as they need.

The Application and Enrichment pages may be scheduled any time after the student has completed the corresponding lesson. You can find helpful teaching tips and the solutions for these pages in the 2012 edition of the *Epsilon Instruction Manual*.

LESSON PRACTICE

Build the problem and then write the correct solution.

1. **Step 1.** Select 15 blocks.

   Step 2. Divide into 5 equal parts.

   Step 3. Count 3 of them.

   —— of ___ is ___

2. **Step 1.** Select 18 blocks.

   Step 2. Divide into 3 equal parts.

   Step 3. Count 2 of them.

   —— of ___ is ___

3. **Step 1.** Select 8 blocks.

   Step 2. Divide into 8 equal parts.

   Step 3. Count 5 of them.

   —— of ___ is ___

Read the problem, build it, and write the answer in the blank.

4. Three fifths of twenty is _____ .

5. Two thirds of six is _____ .

6. One half of eight is _____ .

7. Three fourths of four is _____ .

Solve.

8. $\frac{1}{3}$ of 6 = _____

9. $\frac{1}{4}$ of 8 = _____

10. $\frac{4}{5}$ of 10 = _____

11. $\frac{1}{2}$ of 6 = _____

12. $\frac{2}{3}$ of 12 = _____

13. $\frac{2}{5}$ of 10 = _____

14. Mom bought one dozen eggs. One fourth of them were broken. How many eggs were broken? (If you don't know that one dozen is twelve, this is a good time to learn that fact.)

15. Ten children came to Sam's birthday party. One half of them were boys. How many boys came to the party?

Build the problem and then write the correct solution.

1. **Step 1.** Select 12 blocks.

   Step 2. Divide into 3 equal parts.

   Step 3. Count 1 of them.

   —— of ___ is ___

2. **Step 1.** Select 20 blocks.

   Step 2. Divide into 4 equal parts.

   Step 3. Count 3 of them.

   —— of ___ is ___

3. **Step 1.** Select 18 blocks.

   Step 2. Divide into 9 equal parts.

   Step 3. Count 7 of them.

   —— of ___ is ___

Read the problem, build it, and write the answer in the blank.

4. Two fourths of sixteen is _____ .

5. One fifth of ten is _____ .

6. Five sixths of eighteen is _____ .

7. One third of nine is _____ .

Solve.

8. $\frac{3}{4}$ of 12 = _____

9. $\frac{3}{5}$ of 10 = _____

10. $\frac{1}{2}$ of 4 = _____

11. $\frac{1}{3}$ of 12 = _____

12. $\frac{1}{2}$ of 12 = _____

13. $\frac{7}{8}$ of 16 = _____

14. It rained $\frac{3}{7}$ of the days in the last two weeks. How many days did it rain? (Remember, there are seven days in one week.)

15. Emaleigh planted 20 tulip bulbs in her garden. Four fifths of them came up in the spring. How many tulip plants did she have that spring?

SYSTEMATIC REVIEW

Build the problem and then write the correct solution.

1. **Step 1.** Select 15 blocks.

   Step 2. Divide into 5 equal parts.

   Step 3. Count 2 of them.

   —— of ___ is ___

2. **Step 1.** Select 9 blocks.

   Step 2. Divide into 3 equal parts.

   Step 3. Count 1 of them.

   —— of ___ is ___

Read the problem, build it, and write the answer in the blank.

3. Five ninths of eighteen is _____ .

4. Two fifths of ten is _____ .

Solve. Some of these may be too large to build.

5. $\frac{3}{4}$ of 8 = ____   6. $\frac{5}{6}$ of 30 = ____

7. $\frac{7}{8}$ of 48 = ____   8. $\frac{3}{6}$ of 24 = ____

9. $\frac{4}{7}$ of 35 = ____   10. $\frac{3}{5}$ of 10 = ____

QUICK REVIEW

The perimeter of a rectangle is the distance around the rectangle. It is found by adding the lengths of the four sides. The symbol " stands for *inch*, and the symbol ' stands for *foot* or *feet*.

Find the perimeter of each rectangle. The first one has been done for you.

11.　3"　　5 in + 3 in + 5 in + 3 in = 16 in

　　　5"

　　　　　　　　　　　　　　　　P = <u>16 in</u>

12.　10'　　　　　　　　　　P = _____

　　20'

13.　7"　　　　　　　　　　P = _____

　　15"

14. It took Frank $\frac{2}{3}$ of an hour to walk home from school. How many minutes was that? (There are 60 minutes in one hour. Please memorize this fact.)

15. Three fifths of all the students in Mr. Fulton's class are girls. There are 25 students in the class. How many are girls?

SYSTEMATIC REVIEW

Build the problem and then write the correct solution.

1. **Step 1.** Select 20 blocks.

   Step 2. Divide into 5 equal parts.

   Step 3. Count 4 of them.

   —— of ____ is ____

2. **Step 1.** Select 6 blocks.

   Step 2. Divide into 2 equal parts.

   Step 3. Count 1 of them.

   —— of ____ is ____

Read the problem, build it, and write the answer in the blank.

3. Five fifths of ten is _____ .

4. Three fourths of twenty is _____ .

Solve. Some of these may be too large to build.

5. $\frac{5}{6}$ of 24 = ____

6. $\frac{1}{7}$ of 56 = ____

7. $\frac{2}{3}$ of 12 = ____

8. $\frac{2}{5}$ of 10 = ____

9. $\frac{1}{4}$ of 8 = ____

10. $\frac{3}{8}$ of 32 = ____

Find the perimeter of each rectangle.

11. 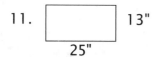 13"

P = _____

12.  16'

P = _____

13. 39"

P = _____

14. Daniel earned $16 and put half of it in the bank. How much money did he put in the bank?

15. Chad wants to put a fence around his garden. The garden is a rectangle that measures 20 feet by 30 feet. How many feet of fence must Chad buy?

16. A bull came into Patty's china shop and broke $\frac{4}{7}$ of the plates. Patty had 56 plates to begin with. How many were broken?

17. How many feet of trim are needed to go around a rectangular window that measures 6 feet by 11 feet?

18. Eighteen students tried out for the school play, but only $\frac{7}{9}$ of them were chosen. How many students were chosen for the school play?

SYSTEMATIC REVIEW

Build the problem and then write the correct solution.

1. **Step 1.** Select 24 blocks.

   Step 2. Divide into 6 equal parts.

   Step 3. Count 3 of them.

   —— of ___ is ___

2. **Step 1.** Select 8 blocks.

   Step 2. Divide into 4 equal parts.

   Step 3. Count 2 of them.

   —— of ___ is ___

Read the problem, build it, and write the answer in the blank.

3. Two thirds of twenty-one is _____ .

4. One third of six is _____ .

Solve. Some of these may be too large to build.

5. $\frac{1}{3}$ of 12 = ____   6. $\frac{3}{4}$ of 12 = ____

7. $\frac{1}{5}$ of 25 = ____   8. $\frac{5}{6}$ of 42 = ____

9. $\frac{1}{9}$ of 54 = ____   10. $\frac{5}{8}$ of 16 = ____

Find the perimeter of each rectangle.

11.   P = _____

12.   P = _____

13.   P = _____

14. There were 36 candies in the bag. Kelsey ate $\frac{5}{9}$ of them. How many candies did she eat?

15. Briley ate $\frac{4}{9}$ of the candies in the bag. (See #14.) How many candies did Briley eat? What is the total number of candies that Kelsey and Briley ate?

16. Stacia earned $12 on Monday, $20 on Tuesday, and $8.50 on Wednesday. How much did she earn during those three days? (When adding or subtracting money, remember to keep the decimal points lined up and, in your answer, put a decimal point directly under them.)

17. Stacia gave $14.50 to her sister Jenna. How much money does Stacia have left from what she earned the last three days? (See #16.)

18. Marissa goes to school $\frac{3}{4}$ of the year (12 months). For how many months does Marissa go to school each year?

To the teacher: Word problems can be challenging. There are tips and strategies for solving word problems in lesson 1 in the instruction manual. We suggest you read them with your student and then let the student use the information to solve the crossword puzzle. Give as much help as is needed.

Across

1. When acting out word problems, don't be afraid to use a little _____ .

4. Use _____ to check your answer.

6. Use the blocks or overlays to _____ a problem.

8. Every word problem tells a _____ .

9. Word problems require both reading skills and _____ skills.

10. Making a _____ can help you picture a problem.

Down

2. The word _____ often means that you should add.

3. Relate word problems to _____ life.

5. It is good practice to _____ your own word problems.

7. _____ words can be useful clues.

In this lesson, you learned how to solve problems that ask you to find the *fraction of a number*. You may have to go to a different part of the problem to find what the number actually represents. Here are some words that you may see used in addition and subtraction problems.

Total – addition
Altogether – addition
Have left or *are left* – subtraction

Underline the important words or phrases and solve the word problems. Not all of these involve fractions. The first one has been done for you.

1. <u>Fifteen boxes</u> needed to be packed. If Zarah packed $\frac{2}{3}$ <u>of them</u>, how many boxes did she pack?

   $\frac{2}{3}$ of 15 boxes = 10 boxes

2. We bought six pies for the family dinner. If four pies have been eaten, how many pies are left?

3. I earned $25 last week and $42 this week. What is the total amount I earned?

Now that you are warmed up, here are some problems with more than one step. Look for key words or phrases and underline them. If the problem involves "fraction of a number" and the number is in a different part of the problem, draw an arrow as in #1 above. Decide what operation to do first and then solve the problem.

4. There were 12 cookies in the bag. Tom ate $\frac{1}{3}$ of the cookies on Monday and $\frac{1}{4}$ of the cookies on Tuesday. How many cookies did Tom eat altogether?

5. Michael planted seven fruit trees in his yard, but three of them died. Of the trees that were left, one half of them were apple trees. How many apple trees does Michael have growing in his yard?

6. Gabriel has $20 in one pocket and $16 in his other pocket. He spent $\frac{3}{4}$ of the total amount of money in his pockets at the bookstore. How much money does Gabriel have left?

Write in the correct numerator and denominator using both symbols and words. The first one has been done for you.

1.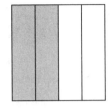

$$\frac{2}{4}$$

two fourths

2.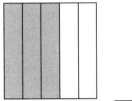

———

————

3.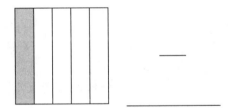

———

————

4.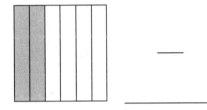

———

————

5.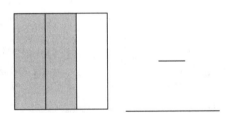

———

————

6.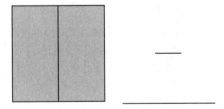

———

————

7.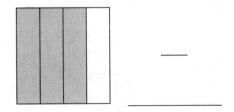

———

————

8.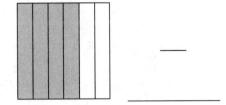

———

————

Build the fraction and then draw it in the square. Say it and write it with words. The first one has been done for you.

9.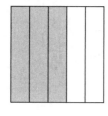

$$\frac{3}{5}$$

three fifths

10.

$$\frac{2}{4}$$

_____

11.

$$\frac{1}{3}$$

_____

12.

$$\frac{2}{2}$$

_____

Build the fraction and then draw it in the square. Say it and write it with numerals. The first one has been done for you.

13.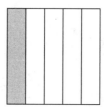

$$\frac{1}{5}$$

one fifth

14.

___

two thirds

15.

___

one fourth

16.

___

three sixths

LESSON PRACTICE

Write the correct numerator and denominator using both symbols and words.

1. 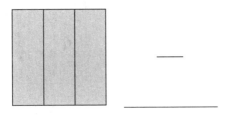 $\dfrac{}{}$

2. 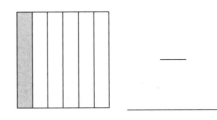 $\dfrac{}{}$

3. 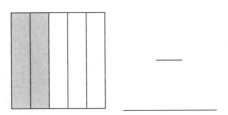 $\dfrac{}{}$

4. 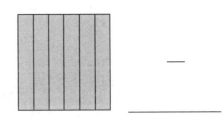 $\dfrac{}{}$

5. 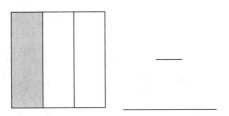 $\dfrac{}{}$

6. 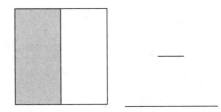 $\dfrac{}{}$

7. 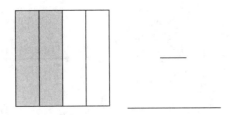 $\dfrac{}{}$

8. 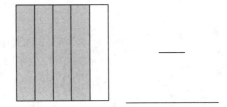 $\dfrac{}{}$

Build the fraction and then draw it in the square. Say it and write it with words.

9. $\dfrac{3}{3}$

_____

10. $\dfrac{1}{6}$

_____

11. $\dfrac{1}{2}$

_____

12. $\dfrac{4}{6}$

_____

Build the fraction and then draw it in the square. Say it and write it with numerals.

13.   ——

three fifths

14.   ——

four fourths

15.   ——

four fifths

16.   ——

five sixths

LESSON PRACTICE

Write the correct numerator and denominator using both symbols and words.

1.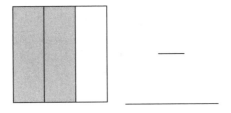
   $$\frac{\quad}{\quad}$$

2.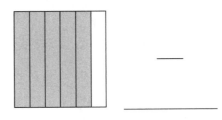
   $$\frac{\quad}{\quad}$$

3.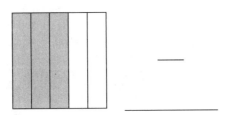
   $$\frac{\quad}{\quad}$$

4.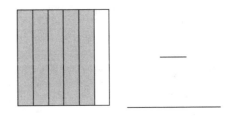
   $$\frac{\quad}{\quad}$$

5.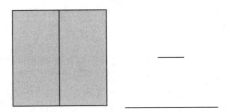
   $$\frac{\quad}{\quad}$$

6.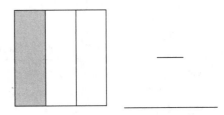
   $$\frac{\quad}{\quad}$$

7.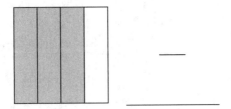
   $$\frac{\quad}{\quad}$$

8.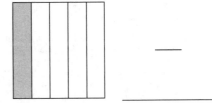
   $$\frac{\quad}{\quad}$$

Build the fraction and then draw it in the square. Say it and write it with words.

9.
$$\frac{2}{6}$$

_____

10.
$$\frac{1}{4}$$

_____

11.
$$\frac{2}{3}$$

_____

12.
$$\frac{4}{5}$$

_____

Build the fraction and then draw it in the square. Say it and write it with numerals.

13.
———

one fifth

14.
———

two fourths

15.
———

one third

16.
———

six sixths

SYSTEMATIC REVIEW

Write the correct numerator and denominator using both symbols and words.

1.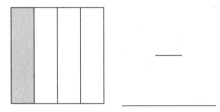

 ____
 _____

2.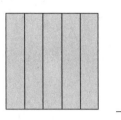

 ____
 _____

3.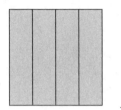

 ____
 _____

Build the fraction and then draw it in the square. Say it and write it using numerals or words.

4.

 $\dfrac{3}{6}$

 _____

5.

 $\dfrac{1}{5}$

 _____

6.

 ____

 five sixths

Solve.

7. $\dfrac{2}{3}$ of 24 = ____

8. $\dfrac{3}{4}$ of 40 = ____

9. $\dfrac{1}{2}$ of 34 = ____

QUICK REVIEW

A square is a special kind of rectangle with all four sides the same length. Its perimeter is also found by adding the sides.

Find the perimeter of each square. The first one has been done for you.

10. 3"   3 in + 3 in + 3 in + 3 in = 12 in   P = __12 in__

3"

11. 15'   P = _____

15'

12. Kayla needs to buy trim to go around the edges of two pillows she is making. One pillow is a rectangle with sides that measure 12 inches and 24 inches. The other pillow is a square that measures 10 inches on a side. How many inches of trim does Kayla need?

13. Eighteen players tried out for the team, but only $\frac{5}{9}$ of them were chosen. How many players were chosen?

14. Lauren planted 12 forsythia bushes. If $\frac{3}{4}$ of them lived, how many bushes did she have?

Write the correct numerator and denominator using both symbols and words.

1. 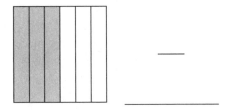 ___

___

2. 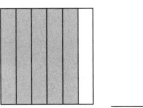 ___

___

3. 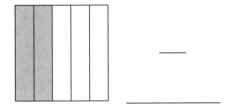 ___

___

Build the fraction and then draw it in the square. Say it and write it using numerals or words.

4. ___

one sixth

5. ___

two thirds

6.  $\dfrac{6}{6}$

___

Solve.

7. $\dfrac{1}{6}$ of 36 = ___

8. $\dfrac{4}{9}$ of 72 = ___

9. $\dfrac{2}{3}$ of 39 = ___

Find the perimeter of each square or rectangle.

10. [□ 17 yd × 4 yd]   4 yd   (yd stands for yards)   P = _____

11. [□ 34' × 34'] 34'   P = _____
     34'

12. Jared has a square garden that measures 18 yards on a side. It has a fence all around the perimeter. Jared wants to replace 10 yards of fence with a hedge. How many yards of fence will be left?

13. In Sally's part of the country it is warm enough to go swimming outdoors $\frac{1}{4}$ of the year. How many months is that?

14. Jayne can go swimming outdoors two thirds of the year. How many warm months is that?

15. The zoo has 35 camels. Five sevenths of them are Bactrian (two-humped) camels. How many Bactrian camels are in the zoo?

Write the correct numerator and denominator using both symbols and words.

1. 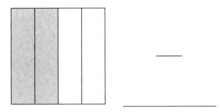 $\dfrac{\rule{1cm}{0.4pt}}{\rule{2cm}{0.4pt}}$

2. 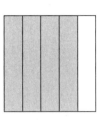 $\dfrac{\rule{1cm}{0.4pt}}{\rule{2cm}{0.4pt}}$

3. 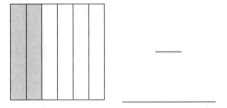 $\dfrac{\rule{1cm}{0.4pt}}{\rule{2cm}{0.4pt}}$

Build the fraction and then draw it in the square. Say it and write it using numerals or words.

4. $\dfrac{\rule{1cm}{0.4pt}}{\text{three fifths}}$

5. $\dfrac{\rule{1cm}{0.4pt}}{\text{two sixths}}$

6. $\dfrac{3}{4}$ $\rule{2cm}{0.4pt}$

Solve.

7. $\dfrac{1}{2}$ of 80 = ____

8. $\dfrac{3}{5}$ of 55 = ____

9. $\dfrac{3}{8}$ of 56 = ____

Find the perimeter of each square or rectangle.

10. [rectangle] 21 yd   P = _____

         46 yd

11. [square] 13'   P = _____

    13'

12. Joanne bought a dozen (12) eggs, but $\frac{1}{6}$ of them were broken. How many eggs were broken?

13. Twenty people came to the party. Three fifths of them went home happy because they won prizes. How many people won prizes?

14. The other $\frac{2}{5}$ of the people at the party were a little disappointed because they didn't win prizes, but they were good sports. How many people were good sports?

15. Jordan earned $10.50 doing chores. He put the money in his wallet along with the $7.25 he already had there. How much money does Jordan now have in his wallet?

In the lessons, a fraction of one is shown using a square. The square makes it easy to show how addition and other operations work with fractions, but one of any shape may be divided into parts. In order to label the parts with a fraction, all of the parts must be the same size.

Follow the directions.

1. Color or shade $\frac{1}{2}$ of one triangle.

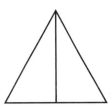

2. Color or shade $\frac{1}{4}$ of one circle.

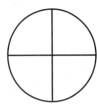

3. Color or shade $\frac{5}{6}$ of one hexagon.

4. Color or shade $\frac{1}{3}$ of one rectangle.

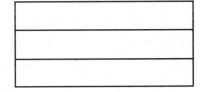

5. Color or shade $\frac{3}{8}$ of one octagon.

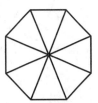

6. Color or shade $\frac{1}{2}$ of one triangle.

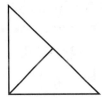

Remember that drawing is a tool for solving word problems. The drawings for these questions have been done for you. Use them to help you "see" the answer.

Follow the directions.

7. Jill had 10 eggs, but she broke $\frac{2}{5}$ of them. Draw lines to divide the eggs into five equal groups. Put Xs on the eggs she broke. Color or shade the eggs that are left.

    How many eggs are broken? _____

    How many eggs are left? _____

8. Fred had eight boxes of books in his room. He decided to donate $\frac{1}{4}$ of the boxes to his local library. Draw lines to divide the boxes into four equal groups. Put Xs on the boxes of books that he will donate. Color or shade the boxes that will be left.

    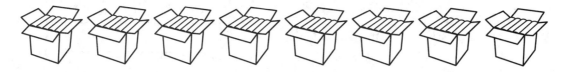

    How many boxes will be donated? _____

    How many boxes will be left? _____

9. Twelve pieces of pie were sitting on the counter. Tom's family ate $\frac{2}{3}$ of the pieces of pie. Draw lines to divide the pieces into three equal groups. Put Xs on the pieces that were eaten. Color or shade the pieces that were left.

    How many pieces of pie were eaten? _____

    How many pieces were left? _____

LESSON PRACTICE

Build and then write your answer using words and symbols.

1.

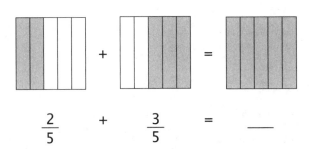

$$\frac{2}{5} \quad + \quad \frac{3}{5} \quad = \quad \underline{}$$

2.

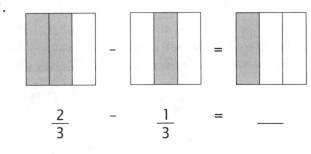

$$\frac{2}{3} \quad - \quad \frac{1}{3} \quad = \quad \underline{}$$

Build, shade, and write your answer using words and symbols.

3.

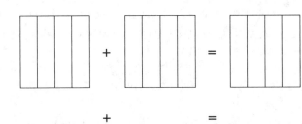

$$\underline{} \quad + \quad \underline{} \quad = \quad \underline{}$$

One fourth plus two fourths equals _____ .

4.

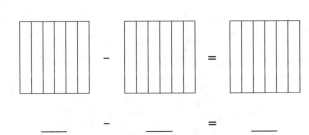

$$\underline{} \quad - \quad \underline{} \quad = \quad \underline{}$$

Three sixths minus two sixths equals _____ .

Build and then solve each problem.

5. $\dfrac{1}{6} + \dfrac{3}{6} = $ ——

6. $\dfrac{1}{5} + \dfrac{2}{5} = $ ——

7. $\dfrac{4}{6} + \dfrac{1}{6} = $ ——

8. $\dfrac{4}{6} - \dfrac{1}{6} = $ ——

9. $\dfrac{3}{4} - \dfrac{2}{4} = $ ——

10. $\dfrac{3}{6} - \dfrac{2}{6} = $ ——

11. Sharlene drew $\frac{1}{4}$ of the letters she needed for her poster. She took a break and then drew $\frac{2}{4}$ more of the letters. What part of the needed letters has she drawn?

12. Teresa read $\frac{1}{5}$ of the book before lunch and $\frac{3}{5}$ of it after lunch. What part of the book has she read?

13. Alexa has done $\frac{2}{6}$ of the job her mother gave her to do. Remember that $\frac{6}{6}$ is the whole job. Subtract to find out what part of the job Alexa has left to complete.

14. Four fifths of the class was present today. One fifth of the total number of students went home early. What part of the total number of students was left in school?

Build and then write your answer using words and symbols.

1.

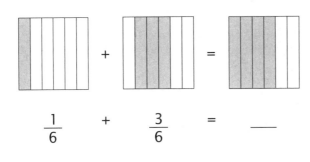

$$\frac{1}{6} \quad + \quad \frac{3}{6} \quad = \quad \underline{}$$

2.

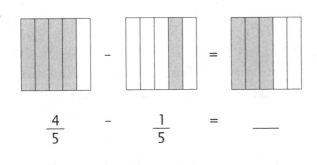

$$\frac{4}{5} \quad - \quad \frac{1}{5} \quad = \quad \underline{}$$

Build, shade, and write each fraction using words. Then express your answer in symbols.

3.

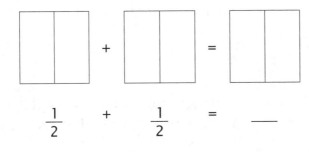

$$\frac{1}{2} \quad + \quad \frac{1}{2} \quad = \quad \underline{}$$

4.

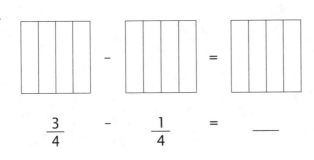

$$\frac{3}{4} \quad - \quad \frac{1}{4} \quad = \quad \underline{}$$

Build and solve each problem.

5. $\dfrac{2}{4} + \dfrac{2}{4} =$ _____

6. $\dfrac{3}{6} + \dfrac{2}{6} =$ _____

7. $\dfrac{1}{5} + \dfrac{3}{5} =$ _____

8. $\dfrac{5}{6} - \dfrac{1}{6} =$ _____

9. $\dfrac{3}{4} - \dfrac{1}{4} =$ _____

10. $\dfrac{2}{3} - \dfrac{1}{3} =$ _____

11. The race cars did $\frac{1}{4}$ of the laps before the red flag went out. When they started racing again, they did $\frac{1}{4}$ more of the laps before the race was canceled because of rain. What part of the laps was completed?

12. Kerri gave one tenth of her pay to disaster relief. What part of her pay was left? Remember that you can write one whole as ten tenths.

13. June drove $\frac{2}{5}$ of her trip on the first day and another $\frac{2}{5}$ of the trip on the second day. What part of the trip has been completed? What part of the trip is left?

14. Terry completed $\frac{2}{6}$ of the job. Lindsay took over for Terry, and when she was finished, $\frac{5}{6}$ of the job was completed. What part of the job did Lindsay do?

LESSON PRACTICE

Build, shade, and then write your answer using words and symbols.

1.

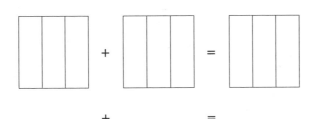

___   +   ___   =   ___

One third plus two thirds equals _____ .

2.

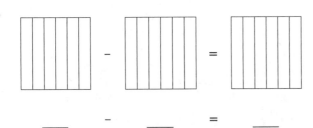

___   -   ___   =   ___

Six sixths minus three sixths equals _____ .

Build, shade, and write each fraction using words. Then express your answer in symbols.

3.

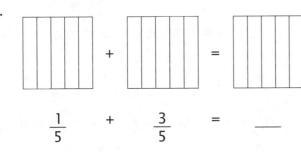

$\frac{1}{5}$ + $\frac{3}{5}$ =   ___

_____

4.

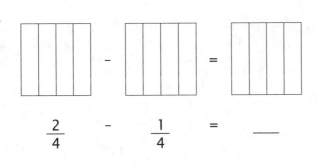

$\frac{2}{4}$ - $\frac{1}{4}$ =   ___

_____

Add or subtract. Some problems cannot be built with the overlays. The first one has been done for you.

5. $\dfrac{4}{9} + \dfrac{2}{9} = \dfrac{6}{9}$

6. $\dfrac{2}{5} + \dfrac{3}{5} = $ ____

7. $\dfrac{6}{7} + \dfrac{1}{7} = $ ____

8. $\dfrac{2}{6} - \dfrac{1}{6} = $ ____

9. $\dfrac{5}{8} - \dfrac{4}{8} = $ ____

10. $\dfrac{6}{10} - \dfrac{1}{10} = $ ____

11. Mom put $\frac{3}{8}$ of a teaspoon of salt in the stew. After tasting it, she added $\frac{1}{8}$ of a teaspoon more salt. How much salt is in the stew?

12. Greg ate $\frac{3}{12}$ of the pizza, and Julia ate $\frac{2}{12}$ of it. What part of the pizza has been eaten?

13. Since $\frac{12}{12}$ is a whole pizza, what part of the whole pizza in #12 was left when Greg and Julia were finished eating?

14. When Tara stopped her bus, $\frac{9}{10}$ of the bus was full. At the stop, she dropped off $\frac{5}{10}$ of the bus load of passengers. How full was the bus after the stop?

SYSTEMATIC REVIEW

Build, shade, and then write your answer using words and symbols.

1.

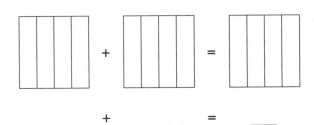

___ + ___ = ___

One fourth plus one fourth equals _____ .

2.

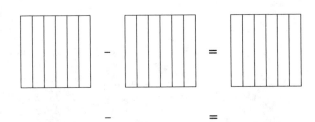

___ - ___ = ___

Five sixths minus one sixth equals _____ .

Add or subtract.

3. $\dfrac{1}{2} + \dfrac{1}{2} =$ ___

4. $\dfrac{3}{10} + \dfrac{6}{10} =$ ___

5. $\dfrac{1}{6} + \dfrac{3}{6} =$ ___

6. $\dfrac{4}{6} - \dfrac{1}{6} =$ ___

7. $\dfrac{3}{4} - \dfrac{2}{4} =$ ___

8. $\dfrac{5}{9} - \dfrac{3}{9} =$ ___

Solve.

9. $\dfrac{4}{5}$ of 10 = ___

10. $\dfrac{5}{8}$ of 40 = ___

11. $\dfrac{2}{7}$ of 49 = ___

QUICK REVIEW

A triangle has three sides. The perimeter is found by adding the lengths of all the sides.

Find the perimeter of each triangle. The first one has been done for you.

12. 4" 4" 5"   4 in + 4 in + 5 in = 13 in   P = __13 in__

13. 3' 5' 4'   P = _____

14. 78' 39' 54'   P = _____

15. Sally ate $\frac{4}{9}$ of her box of chocolates one afternoon. She was getting a bit full, so she let Taylor eat $\frac{1}{9}$ of a box. What part of the chocolates has been eaten?

16. Steve corrected $\frac{3}{8}$ of the math problems. Ted corrected $\frac{2}{8}$ of them. What part of the problems has been corrected? What part of the problems remains to be corrected?

17. Sixteen people showed up for the meeting, but one half of them were late. How many people were late?

18. Brian read $\frac{5}{6}$ of a book that has 24 pages. How many pages has he read?

SYSTEMATIC REVIEW

Build, shade, and write each fraction with words. Then express your answer in symbols.

1.

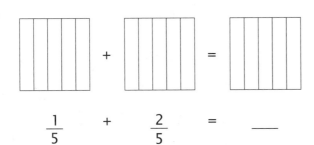

$$\frac{1}{5} \quad + \quad \frac{2}{5} \quad = \quad \underline{\quad}$$

2.

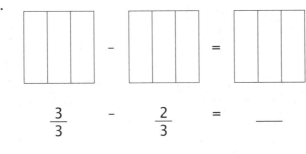

$$\frac{3}{3} \quad - \quad \frac{2}{3} \quad = \quad \underline{\quad}$$

Add or subtract.

3. $\frac{1}{5} + \frac{1}{5} = \underline{\quad}$

4. $\frac{2}{4} + \frac{2}{4} = \underline{\quad}$

5. $\frac{2}{8} + \frac{5}{8} = \underline{\quad}$

6. $\frac{4}{5} - \frac{1}{5} = \underline{\quad}$

7. $\frac{4}{6} - \frac{2}{6} = \underline{\quad}$

8. $\frac{9}{10} - \frac{4}{10} = \underline{\quad}$

Solve.

9. $\frac{3}{4}$ of 12 = \underline{\quad}

10. $\frac{2}{5}$ of 40 = \underline{\quad}

11. $\frac{5}{7}$ of 14 = \underline{\quad}

Find the perimeter of each shape.

12. 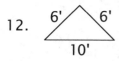   P = _____

13. 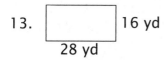   P = _____

14. ☐ 17"   P = _____

    17"

15. Shelley spent $\frac{3}{8}$ of the day collecting shells and $\frac{4}{8}$ of the day selling them. What part of the day did Shelley spend on her project?

16. Riley bought a can of mixed nuts. Two fourths of the nuts were peanuts, and one fourth of them were almonds. What part of the nuts was peanuts or almonds? What part of the nuts was not peanuts or almonds?

17. Emma is buying edging for two triangular flower beds by her front door. Each of the beds has sides that measure 6 feet, 8 feet, and 12 feet. How many feet of edging must Emma buy in all?

18. Mom bought a bag of apples. Nathan ate two fifths of them, and Michael ate one fifth of them. What part of the apples has been eaten?

19. There were 10 apples in the bag when Mom brought it home. (See #18.) How many apples were left when Nathan and Michael were finished eating?

20. Allie spent $3.15 on a sandwich, $0.95 on a soda, and $1.25 for chips. How much money did she spend on lunch?

SYSTEMATIC REVIEW

3F

Build, shade, and write your answer using words and symbols.

1.

___ + ___ = ___

Three sizths plus two sixths equals_____.

2.

___ - ___ = ___

Four fourths minus two fourths equals_____.

Add or subtract.

3. $\frac{1}{3} + \frac{1}{3} =$ ___   4. $\frac{2}{5} + \frac{1}{5} =$ ___

5. $\frac{3}{11} + \frac{4}{11} =$ ___   6. $\frac{5}{9} - \frac{4}{9} =$ ___

7. $\frac{6}{7} - \frac{1}{7} =$ ___   8. $\frac{3}{3} - \frac{1}{3} =$ ___

Solve.

9. $\frac{3}{7}$ of 28 = ___   10. $\frac{1}{3}$ of 27 = ___

11. $\frac{3}{4}$ of 36 = ___

EPSILON SYSTEMATIC REVIEW 3F

43

Find the perimeter of each shape.

12. 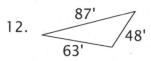 87' 48' 63'   P = _____

13. [ ] 18 yd   P = _____
35 yd

14. [ ] 81"   P = _____
81"

15. Rhonda the snail traveled $\frac{1}{5}$ of a foot and then stopped for a rest. She then traveled as fast as she could for another $\frac{3}{5}$ of a foot. What part of a foot has she traveled?

16. I ordered ten pizzas for my party. Only $\frac{4}{5}$ of them were eaten. How many pizzas were eaten? How many pizzas are left over?

17. Chris ate $\frac{1}{3}$ of his sandwich before the phone rang. How much of his sandwich is left to be eaten?

18. David plans to build a square deck on the back of his house. He will need to build a railing around three sides of the deck. Each side of the square is 16 feet long. How many feet of railing does he need?

19. Laura had $17.45 when she entered the store and $3.19 when she came out. How much had she spent?

20. Anna spent one sixth of the year in New York. How many months did she spend in New York?

In your instruction manual there are mental math problems written with words. Sometimes it is useful to take words and turn them into number problems. Parentheses can help you know what part of the problem to do first and what part to do next.

Example 1
Words: Add three and five and then multiply by six.

First, write 3 + 5 and put parentheses around the whole expression:

$(3 + 5)$

Then multiply by 6. The resulting expression is six times the sum of the numbers in the parentheses.

$6 \times (3 + 5) =$

To find the answer, work inside the parentheses first:

Think "3 + 5 = 8," and then think "6 × 8 = 48."

Write an expression for each sentence. Fill in the blanks without calculating the answer. Then find the exact answer to each problem.

1. Add six and four and then multiply the result by two.

   The answer is _____ times the amount in the parentheses.

2. Add five and two and then multiply the result by eight.

   The answer is _____ times the amount in the parentheses.

3. Subtract four from nine and then multiply the result by three.

   The answer is _____ times the amount in the parentheses.

4. Subtract 345 from 1,268 and then multiply the result by ten.

   The answer is _____ times the amount in the parentheses.

In the first two lessons, the fraction word problems asked for a "fraction of" something. In other words, you found the fractional part of a number. Study the word problems on pages 3A, 3B, and 3C. Notice that the problems use the phrase "fraction of", but there is no specific number given that represents the whole. Instead, you added the parts, and the answer was still a *part* of the *whole*.

Sometimes the *whole* is a number that is given, such as 10 cookies or 15 apples. Other times you may be asked to find the total amount or part of the total amount of something. Study the example. Notice the two different key questions: "What part?" and "How many?"

Example 2
There was a bowl of apples on the table. Ryan ate $\frac{1}{5}$ of them, and Chris ate $\frac{2}{5}$ of them. **What part** of the apples did the boys eat? If there were 15 apples to start with, **how many** apples did the boys eat altogether?

There are two questions, so the problem has two parts.

What part? $\frac{1}{5} + \frac{2}{5} = \frac{3}{5}$, so the boys ate $\frac{3}{5}$ of the apples.
How many? $\frac{3}{5}$ of 15 = 9, so the boys ate 9 apples altogether.

Solve the two-part word problems. Underline the key questions if you wish. Sometimes the number of the whole is given first in the question. Add the fractional parts first to find "what part," and use that answer to find "how many."

5. Sue has 12 seashells. She used $\frac{1}{6}$ of them to make a bracelet and $\frac{3}{6}$ of them to make a necklace.

   What part of the seashells did she use to make jewelry?

   How many seashells did she use to make jewelry?

6. There were eight prizes on the table at the party. Jim won $\frac{1}{4}$ of them, and Bob won $\frac{1}{4}$ of them.

   What part of the prizes were won by Jim and Bob?

   How many prizes were won by Jim and Bob?

7. Steve had 20 math problems to solve. He solved $\frac{2}{5}$ of them yesterday and $\frac{3}{5}$ of them today.

   What part of his math problems has Steve solved so far?

   How many math problems has he solved?

LESSON PRACTICE

Build the equivalent fractions. Draw lines and fill in the blanks to show what you have built. The first one has been done for you.

1.

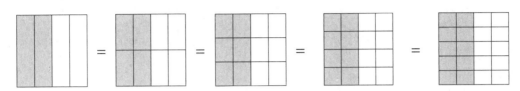

$$\frac{2}{4} = \frac{4}{8} = \frac{6}{12} = \frac{8}{16} = \frac{10}{20}$$

two
fourths

four
eighths

six
twelfths

eight
sixteenths

ten
twentieths

2.

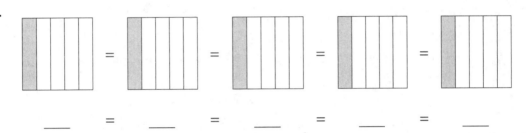

$$\underline{\quad} = \underline{\quad} = \underline{\quad} = \underline{\quad} = \underline{\quad}$$

3.

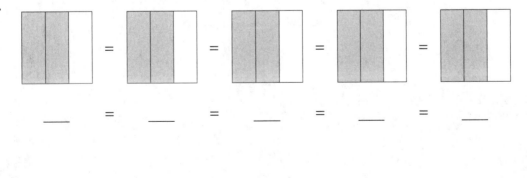

$$\underline{\quad} = \underline{\quad} = \underline{\quad} = \underline{\quad} = \underline{\quad}$$

4.

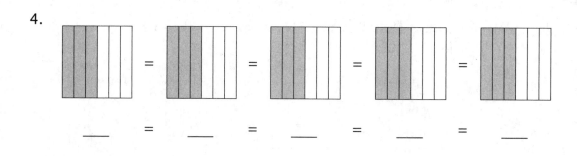

___ = ___ = ___ = ___ = ___

_____ _____ _____ _____ _____

Fill in the missing numbers to make equivalent fractions. See if you can find a pattern in the numbers. Use the overlays if needed. The first one has been done for you.

5. $\dfrac{1}{2} = \dfrac{2}{4} = \dfrac{3}{6} = \dfrac{4}{8}$

6. $\dfrac{3}{5} = \dfrac{}{10} = \dfrac{9}{} = \dfrac{}{20}$

Build the equivalent fractions. Draw lines and fill in the blanks to show what you have built.

1.

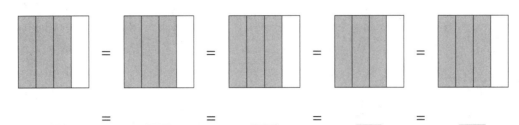

___ = ___ = ___ = ___ = ___

_____

2.

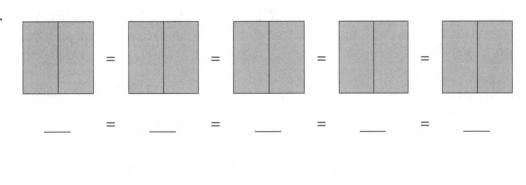

___ = ___ = ___ = ___ = ___

_____

3.

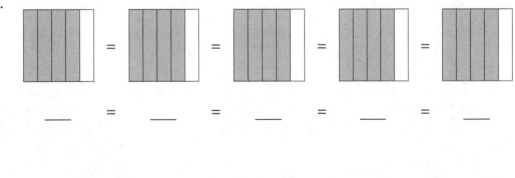

___ = ___ = ___ = ___ = ___

_____

Fill in the missing numbers to make equivalent fractions. Use the overlays if needed. Notice the patterns in the numerator and the denominator.

4. $\dfrac{1}{3} = \dfrac{}{6} = \dfrac{3}{} = \dfrac{}{12}$

5. $\dfrac{2}{4} = \dfrac{}{8} = \dfrac{6}{} = \dfrac{}{16}$

6. $\dfrac{5}{6} = \dfrac{10}{} = \dfrac{}{18} = \dfrac{20}{}$

7. $\dfrac{3}{3} = \dfrac{6}{} = \dfrac{}{9} = \dfrac{12}{}$

8. One third of Bonnie's sheep have black wool. How many sixths is that?

LESSON PRACTICE

Build the equivalent fractions. Draw lines and fill in the blanks to show what you have built.

1.

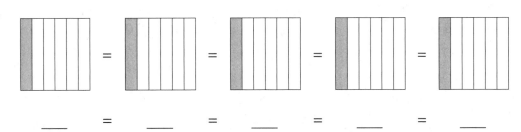

____ = ____ = ____ = ____ = ____

_____ _____ _____ _____ _____

2.

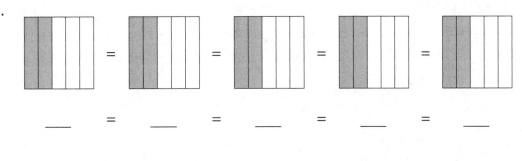

____ = ____ = ____ = ____ = ____

_____ _____ _____ _____ _____

3.

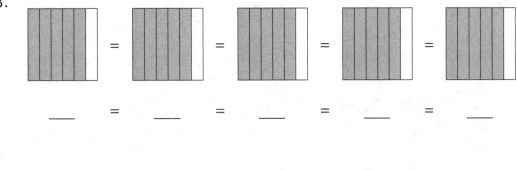

____ = ____ = ____ = ____ = ____

_____ _____ _____ _____ _____

Fill in the missing numbers to make equivalent fractions. Use the overlays if needed. Notice the skip counting patterns in the numerator and the denominator.

4. $\dfrac{3}{4} = \dfrac{}{8} = \dfrac{9}{} = \dfrac{}{16}$

5. $\dfrac{1}{5} = \dfrac{}{10} = \dfrac{3}{} = \dfrac{}{20}$

6. $\dfrac{4}{5} = \dfrac{8}{} = \dfrac{}{15} = \dfrac{16}{}$

7. $\dfrac{2}{3} = \dfrac{4}{} = \dfrac{}{9} = \dfrac{8}{}$

8. Bob read one fourth of his book yesterday. How many eighths of his book did he read?

SYSTEMATIC REVIEW

Build the equivalent fractions. Draw lines and fill in the blanks to show what you have built.

1.

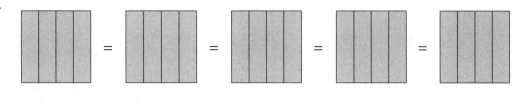

$\dfrac{\quad}{\quad}$ = $\dfrac{\quad}{\quad}$ = $\dfrac{\quad}{\quad}$ = $\dfrac{\quad}{\quad}$ = $\dfrac{\quad}{\quad}$

four
fourths _____ _____ _____ _____

Fill in the missing numbers to make equivalent fractions.

2. $\dfrac{1}{4} = \dfrac{\quad}{\quad} = \dfrac{\quad}{12} = \dfrac{4}{\quad}$

3. $\dfrac{1}{5} = \dfrac{\quad}{10} = \dfrac{3}{\quad} = \dfrac{\quad}{20}$

4. $\dfrac{2}{6} = \dfrac{\quad}{12} = \dfrac{\quad}{\quad} = \dfrac{8}{\quad}$

5. $\dfrac{3}{3} = \dfrac{\quad}{\quad} = \dfrac{9}{\quad} = \dfrac{\quad}{12}$

Add or subtract.

6. $\dfrac{3}{6} + \dfrac{2}{6} = \dfrac{\quad}{\quad}$

7. $\dfrac{5}{7} - \dfrac{3}{7} = \dfrac{\quad}{\quad}$

8. $\dfrac{1}{4} + \dfrac{1}{4} = \dfrac{\quad}{\quad}$

Solve.

9. $\dfrac{1}{6}$ of 36 = ____

10. $\dfrac{5}{8}$ of 32 = ____

11. $\dfrac{3}{9}$ of 81 = ____

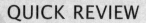

QUICK REVIEW

When multiplying, keep place value in mind. In #12, the "8" goes in the tens column because $2 \times 4 = 8$ is actually $20 \times 4 = 80$. Writing the "0" in the units place is optional.

Multiply. None of these problems require regrouping. The first one has been done for you.

```
12.      3 4
       × 2 1
         3 4
       6 8 0
       7 1 4
```

```
13.      1 6
       × 1 1
```

```
14.    1 2 3
       ×  3 2
```

15. What number should be multiplied by the numerator and denominator of $\frac{1}{3}$ to make the equivalent fraction $\frac{2}{6}$?

16. Danny ate $\frac{1}{2}$ of a pie, and Matt ate $\frac{4}{8}$ of a pie. Did they eat the same amount of pie?

17. Christie has earned $\frac{3}{4}$ of the money she needs for a new game. The game costs $36. How much has she earned?

18. Don suggested the family do $\frac{2}{7}$ of the housekeeping chores in the morning and $\frac{3}{7}$ in the afternoon. If they follow his advice, what part of the chores will be left to complete?

Build the equivalent fractions. Draw lines and fill in the blanks to show what you have built.

1.

___ = ___ = ___ = ___ = ___

_____ _____ _____ _____ _____

Fill in the missing numbers to make equivalent fractions.

2. $\dfrac{4}{6} = \dfrac{\quad}{\quad} = \dfrac{\quad}{18} = \dfrac{16}{\quad}$

3. $\dfrac{2}{5} = \dfrac{\quad}{10} = \dfrac{6}{\quad} = \dfrac{\quad}{20}$

4. $\dfrac{1}{2} = \dfrac{\quad}{\quad} = \dfrac{\quad}{\quad} = \dfrac{4}{8}$

5. $\dfrac{1}{3} = \dfrac{\quad}{\quad} = \dfrac{3}{\quad} = \dfrac{\quad}{12}$

Add or subtract.

6. $\dfrac{3}{4} + \dfrac{1}{4} = \underline{\quad}$

7. $\dfrac{6}{9} - \dfrac{2}{9} = \underline{\quad}$

8. $\dfrac{8}{8} - \dfrac{5}{8} = \underline{\quad}$

Solve.

9. $\dfrac{4}{7}$ of 49 = _____

10. $\dfrac{3}{5}$ of 45 = _____

11. $\dfrac{1}{2}$ of 12 = _____

Multiply.

12. 2 2
    × 1 2

13. 2 3
    × 1 3

14. 4 0 5
    × 1 1

15. Joanne tested $\frac{2}{3}$ of her class on multiplication. How many sixths of the class did she test?

16. There are 18 students in Joanne's class (#15). How many of the students were tested on multiplication?

17. Matt mowed $\frac{1}{2}$ of the lawn. How many fourths did he mow?

18. Everett was $\frac{1}{4}$ of a century (100 years) old when he got married. Since then, he has lived another $\frac{2}{4}$ of a century. For what part of a century has Everett been alive?

19. Grace ordered flowering bushes for her yard. One sixth of the bushes had blue flowers, and the rest of them had red flowers. What part of the bushes had red flowers?

20. Grace ordered 12 bushes in all. (See #19.) How many of each color should she receive?

Build the equivalent fractions. Draw lines and fill in the blanks to show what you have built.

1.

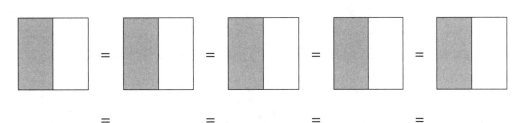

___ = ___ = ___ = ___ = ___

_____ _____ _____ _____ _____

Fill in the missing numbers to make equivalent fractions. You will have to use skip counting or multiplication for #4 and #5, as you do not have overlays for them.

2. $\dfrac{1}{6} = \dfrac{}{} = \dfrac{}{18} = \dfrac{4}{}$

3. $\dfrac{3}{5} = \dfrac{}{} = \dfrac{9}{} = \dfrac{}{20}$

4. $\dfrac{1}{7} = \dfrac{}{} = \dfrac{}{} = \dfrac{4}{28}$

5. $\dfrac{3}{8} = \dfrac{}{} = \dfrac{9}{} = \dfrac{}{32}$

Add or subtract.

6. $\dfrac{2}{3} - \dfrac{1}{3} = \underline{}$

7. $\dfrac{3}{5} + \dfrac{1}{5} = \underline{}$

8. $\dfrac{3}{10} - \dfrac{2}{10} = \underline{}$

Solve.

9. $\dfrac{1}{10}$ of 60 = _____

10. $\dfrac{2}{8}$ of 16 = _____

11. $\dfrac{3}{3}$ of 9 = _____

Multiply.

12. 1 2
     × 1 1

13. 1 4
     × 1 2

14. 2 2 1
     × 4 3

15. Mrs. Anderson was going to make applesauce, but she had to throw away $\frac{1}{7}$ of her apples because they were spoiled. How many fourteenths did she throw away?

16. Mrs. Anderson (#15) started with 14 pounds of apples. How many pounds of apples did she throw away?

17. A baker's dozen is 13. If a baker sold 11 baker's dozens of rolls, how many did he sell?

18. Each side of a square measures four feet in length. What is the perimeter of the square?

19. Caleb drove $\frac{3}{8}$ of the distance to the campground. Then Timothy drove for $\frac{1}{8}$ of the distance. What part of the trip is left to drive?

20. Daniel drove for the rest of the trip in #19. The entire trip was 64 miles. How many miles did Daniel drive?

Follow the directions.

1. Ian had $\frac{1}{2}$ of a cake left over from his party. Shade or color the cake to show what part of a whole cake he had left. Before Ian could eat his cake, his brother Jamie came into the room. Ian wanted to share, so he cut the cake into two pieces that were the same size. Draw a line to show how Ian cut the cake.

2. Did cutting the cake change the amount of the cake in the pan? Write a fraction equivalent to $\frac{1}{2}$ to tell what part of a whole cake is in the pan.

3. Use a fraction to describe Ian's share of the whole cake. What is Jamie's share of the whole cake?

4. Callie had $\frac{3}{4}$ of a whole cake. Shade or color the cake to show what part of a whole cake she had.

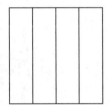

5. Callie thought it would be too much cake to eat alone, so she invited Julia, Renna, and Rachel to help her. Draw lines to show how Callie should cut the cake. Write a fraction equivalent to $\frac{3}{4}$ to tell what part of a whole cake is in the pan.

6. Write a fraction that tells what part of a whole cake each girl should receive.

Here is a fraction word problem based on a circle instead of a square.

7. Mom made a blueberry pie for the family. One half of the pie is left in the pan. Shade or color the pie to show what part is left in the pan.

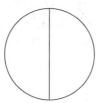

8. Mom said that James and Jeff could have the rest of the pie if they shared it equally. Draw a line to show how much pie each boy will eat. Write a fraction equivalent to $\frac{1}{2}$ to tell what part of the whole blueberry pie is left to be eaten.

9. Write a fraction telling what part of the whole pie each boy will eat. Write another fraction telling what part of the leftover pie each boy will eat.

You know how to add fractions to find a total. Next, start with a fraction that has a numerator greater than one and see how many ways that fraction can be expressed as an addition problem.

Example 1

$\frac{3}{7}$ could be $\frac{1}{7} + \frac{1}{7} + \frac{1}{7}$ or $\frac{2}{7} + \frac{1}{7}$.

Rewrite the following fractions using addition in at least two different ways.

10. $\frac{4}{5}$

11. $\frac{3}{6}$

12. $\frac{7}{8}$

Build the equivalent fractions. Circle the fractions with the same or "common" denominators in each pair and use them to finish the addition problem. In some cases, the small rectangles in the fractions with common denominators may not look the same. Do not be concerned—remember that rectangles with the same area may have different dimensions. The first one has been done for you.

1.

$$\frac{2}{4} = \frac{4}{8} = \boxed{\frac{6}{12}} = \frac{8}{16} = \frac{10}{20}$$

$$\frac{1}{6} = \boxed{\frac{2}{12}} = \frac{3}{18} = \frac{4}{24} = \frac{5}{30}$$

Add. $\frac{6}{12} + \frac{2}{12} = \frac{8}{12}$

2.

$$\underline{\quad} = \underline{\quad} = \underline{\quad} = \underline{\quad} = \underline{\quad}$$

$$\underline{\quad} = \underline{\quad} = \underline{\quad} = \underline{\quad} = \underline{\quad}$$

Add. $\underline{\quad} + \underline{\quad} = \underline{\quad}$

3.

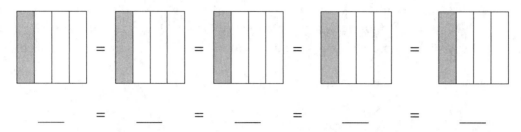

____ = ____ = ____ = ____ = ____

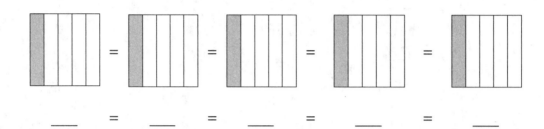

____ = ____ = ____ = ____ = ____

Add. ____ + ____ = ____

4. Use the overlays to build and solve. $\dfrac{1}{2} + \dfrac{1}{6} =$ ____

Build the equivalent fractions. Circle the fractions with the same or "common" denominators in each pair and use them to finish the subtraction problem.

1.

$$\rule{1cm}{0.4pt} = \rule{1cm}{0.4pt} = \rule{1cm}{0.4pt} = \rule{1cm}{0.4pt} = \rule{1cm}{0.4pt}$$

$$\rule{1cm}{0.4pt} = \rule{1cm}{0.4pt} = \rule{1cm}{0.4pt} = \rule{1cm}{0.4pt} = \rule{1cm}{0.4pt}$$

Subtract. $\rule{1cm}{0.4pt} - \rule{1cm}{0.4pt} = \rule{1cm}{0.4pt}$

2.

$$\rule{1cm}{0.4pt} = \rule{1cm}{0.4pt} = \rule{1cm}{0.4pt} = \rule{1cm}{0.4pt} = \rule{1cm}{0.4pt}$$

$$\rule{1cm}{0.4pt} = \rule{1cm}{0.4pt} = \rule{1cm}{0.4pt} = \rule{1cm}{0.4pt} = \rule{1cm}{0.4pt}$$

Subtract. $\rule{1cm}{0.4pt} - \rule{1cm}{0.4pt} = \rule{1cm}{0.4pt}$

3.

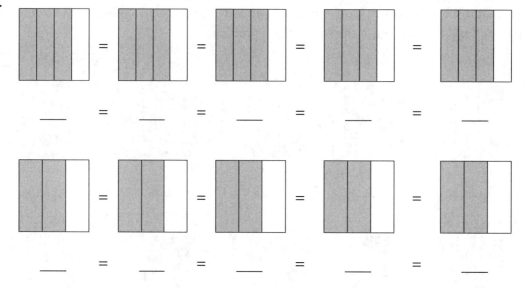

_____ = _____ = _____ = _____ = _____

_____ = _____ = _____ = _____ = _____

Subtract. _____ - _____ = _____

4. Use the overlays to build and solve. $\dfrac{2}{4} - \dfrac{1}{5} =$ _____

Use the overlays to change each fraction to an equivalent fraction and add or subtract. The first one has been done for you.

1.

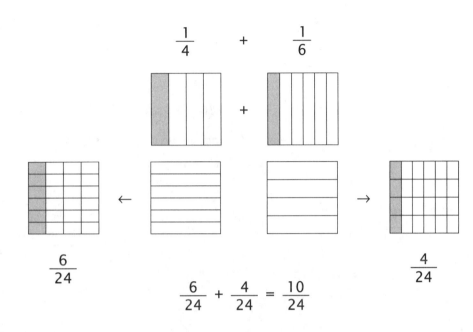

$$\frac{6}{24} + \frac{4}{24} = \frac{10}{24}$$

2.

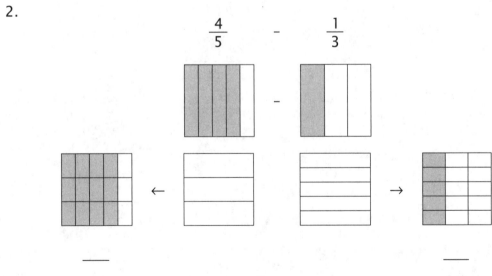

After building these problems, draw lines to show the result of using the overlays and then finish each one.

3.

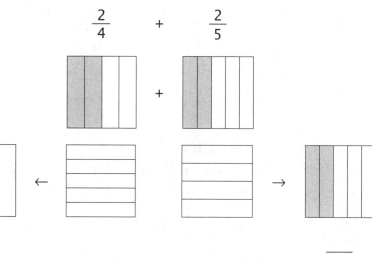

___

___ + ___ = ___

4.

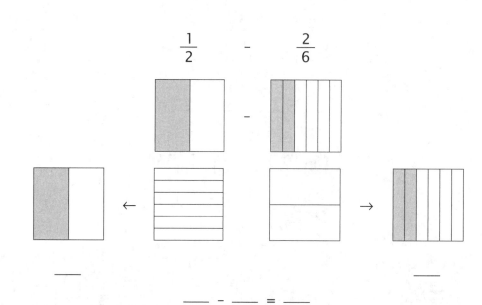

___

___ - ___ = ___

Use the overlays to build and solve.

5. $\dfrac{1}{3} + \dfrac{3}{6} =$ ___

6. $\dfrac{3}{5} - \dfrac{1}{3} =$ ___

7. $\dfrac{2}{4} + \dfrac{1}{5} =$ ___

SYSTEMATIC REVIEW

Use the overlays to find equivalent fractions, using the shortcut as you did on page 5C. Then add or subtract the fractions. It is important to note that using this method will not always give the least common denominator. Do not be concerned with simplifying fractions at this point. Simplifying will be taught in a later lesson.

1. $\dfrac{1}{2} + \dfrac{1}{3} = $ ——

2. $\dfrac{1}{3} + \dfrac{2}{5} = $ ——

3. $\dfrac{3}{6} + \dfrac{1}{4} = $ ——

4. $\dfrac{3}{6} - \dfrac{1}{3} = $ ——

5. $\dfrac{2}{3} - \dfrac{1}{6} = $ ——

6. $\dfrac{1}{5} - \dfrac{1}{6} = $ ——

Fill in the missing numbers in the numerators or denominators to make equivalent fractions.

7. $\dfrac{2}{3} = \dfrac{}{6} = \dfrac{}{9} = \dfrac{8}{}$

8. $\dfrac{4}{7} = \dfrac{}{} = \dfrac{}{} = \dfrac{}{28}$

Solve.

9. $\dfrac{1}{4}$ of 24 = ____

10. $\dfrac{2}{5}$ of 25 = ____

11. $\dfrac{1}{7}$ of 63 = ____

QUICK REVIEW

Rounding is used in estimation, which will be reviewed in another lesson. When rounding, look at the digit to the right of the place you want to round. If that digit is zero through four, the place you are rounding to stays the same. If the digit to the right is five through nine, the place you are rounding to increases by one. Study the examples.

Example 1
Rounding to the nearest ten:

$2\underline{3} \rightarrow 20$

$4\underline{6} \rightarrow 50$

Example 2
Rounding to the nearest hundred:

$1\underline{4}8 \rightarrow 100$

$5\underline{5}1 \rightarrow 600$

Round to the nearest ten.

12. 31 ____

13. 78 ____

Round to the nearest hundred.

14. 415 ____

15. 650 ____

16. A grocer sold 60 Thanksgiving turkeys. Five sixths of them were frozen. How many frozen turkeys were sold?

17. Jason ate $\frac{2}{3}$ of a pizza, and Pat ate $\frac{1}{6}$ of a pizza. What part of a pizza was eaten? Use the overlays to set up this addition problem.

18. How many months are in 31 years?

Use the overlays to find equivalent fractions. Then add or subtract the fractions.

1. $\dfrac{1}{5} + \dfrac{1}{6} =$ ____

2. $\dfrac{1}{3} + \dfrac{3}{5} =$ ____

3. $\dfrac{1}{4} + \dfrac{1}{5} =$ ____

4. $\dfrac{5}{6} - \dfrac{1}{5} =$ ____

5. $\dfrac{4}{5} - \dfrac{3}{4} =$ ____

6. $\dfrac{1}{3} - \dfrac{1}{6} =$ ____

Fill in the missing numbers in the numerators or denominators to make equivalent fractions.

7. $\dfrac{1}{2} =$ ____ $= \dfrac{}{6} = \dfrac{4}{}$

8. $\dfrac{3}{6} =$ ____ $=$ ____ $= \dfrac{}{24}$

Solve.

9. $\dfrac{1}{3}$ of 18 = ____

10. $\dfrac{3}{7}$ of 42 = ____

11. $\dfrac{2}{9}$ of 9 = ____

Round to the nearest ten.

12. 49 ____   13. 62 ____

Round to the nearest hundred.

14. 835 ____   15. 119 ____

Always think carefully about answers to word problems to be sure they are sensible. Use the fraction overlays as needed to help you "see" the answer.

16. Mindy ran for one third of a mile and walked for one half of a mile. How far did she travel altogether?

17. Two fifths of the children brought pears in their lunch boxes, and one fourth of them brought apples. What part of the children brought either pears or apples in their lunch boxes?

18. The total number of children in #17 was twenty. Find out the number of children who brought pears and the number of children who brought apples.

19. One afternoon 221 prairie dogs could be seen. Each prairie dog had four dirty paws. How many dirty paws could be seen?

20. The three sides of a triangle measured 8 feet, 13 feet, and 19 feet. What was the perimeter of the triangle?

Use the overlays to find equivalent fractions. Then add or subtract the fractions.

1. $\dfrac{2}{4} + \dfrac{2}{6} = $ ——

2. $\dfrac{1}{4} + \dfrac{3}{5} = $ ——

3. $\dfrac{1}{3} + \dfrac{2}{6} = $ ——

4. $\dfrac{1}{2} - \dfrac{1}{3} = $ ——

5. $\dfrac{2}{5} - \dfrac{1}{4} = $ ——

6. $\dfrac{5}{6} - \dfrac{1}{3} = $ ——

Fill in the missing numbers in the numerators or denominators to make equivalent fractions.

7. $\dfrac{3}{4} = \dfrac{\quad}{\quad} = \dfrac{\quad}{12} = \dfrac{12}{\quad}$

8. $\dfrac{5}{9} = \dfrac{\quad}{\quad} = \dfrac{\quad}{\quad} = \dfrac{20}{36}$

Solve.

9. $\dfrac{2}{5}$ of 25 = ____

10. $\dfrac{2}{3}$ of 30 = ____

11. $\dfrac{1}{8}$ of 56 = ____

Round to the nearest ten.

12. 55 ___

13. 11 ___

Round to the nearest hundred.

14. 109 ___

15. 360 ___

Continue to think carefully about answers to word problems to be sure they are sensible. Use the fraction overlays as needed to help you "see" the answer.

16. The scarf that Miranda is knitting was $\frac{1}{2}$ of a yard long by lunch time. An hour after lunch, it was $\frac{3}{4}$ of a yard long. What part of a yard has Miranda knitted since lunch? Use the fraction overlays to check the reasonableness of your answer.

17. Eighty people marched in the parade. Three eighths of them carried flags. How many people carried flags?

18. Janna bought three birthday gifts for members of her family. The gifts cost $15.34, $19.99, and $16.50. What is the total amount she spent on gifts?

19. A rectangle has a length of 50 feet and a width of 30 feet. What is the perimeter of the rectangle?

20. Each bench seats 12 people. How many people can sit on 24 benches?

Here are some hints to help you decide whether to add or subtract.

Hint #1: Decide whether there will be more or less of something when the action is finished. Add to get more and subtract to get less.

Hint #2: Substitute whole numbers for the fractions when you read the problem. Once you have decided on the operation, add or subtract the fractions to find the answer.

Hint #3: "Altogether," "another," "total," and "in all" may indicate addition. "Has left," "difference," and "how much remains" may indicate subtraction. These key words can help, but you still need to read through the problem for meaning.

Read each word problem. Underline any key words that may help you know what operation is needed to solve the problem. Circle the correct operation.

1. Conner rode his horse for a fraction of a mile and then rode his bike for another fraction of a mile. How far did he ride in all?

   add   subtract

2. One fraction of my garden was planted with vegetables. The rest of the garden is not planted yet. What part of the garden is left to be planted?

   add   subtract

3. Cody caught a fish that weighed a fraction of a pound. Dennis caught a fish that weighed a different fraction of a pound. What is the difference in the weights of their fish?

   add   subtract

4. Kristi had some jelly beans. One part of the jelly beans was red, and one part was green. What part of the jelly beans was either red or green?

   add   subtract

5. Christopher read a fraction of the pages in his book. What part of his book is left to be read?

   add   subtract

When adding and subtracting fractions, notice that sometimes you are working with fractions of one whole and sometimes with fractions of a number. A drawing can help with both types of problems, but the drawings will look different.

Here are some of the problems from the last page with numbers added. Use drawings as suggested to show the solutions to each of the word problems.

6. Conner rode his horse for $\frac{1}{4}$ of a mile and then rode his bike for another $\frac{1}{4}$ of a mile. How far did he ride in all? Shade the drawing to show how far he rode. Write and solve the problem.

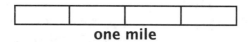
one mile

7. One third of my garden was planted with vegetables. The rest of the garden is not planted yet. What part of the garden is left to be planted? Make a drawing to represent the garden. Write and solve the problem.

8. Cody caught a fish that weighed $\frac{3}{4}$ of a pound. Dennis caught a fish that weighed $\frac{1}{2}$ of a pound. What was the difference in the weights of their fish? Make a separate square drawing for each fish. Let each square represent one pound. Write and solve the problem.

9. Kristi had 12 jelly beans. One third of the jelly beans were red, and one sixth were green. What part of the jelly beans were either red or green? This problem is about the fraction of a number. First, draw 12 jelly beans. Color one third of them red and one sixth of them green. Write and solve the problem.

10. The answer to #9 tells what part of the jelly beans are either red or green. Can you use the fraction to find the actual number of jelly beans that are either red or green? Does the answer agree with the answer you get by counting the colored jelly beans in your picture?

LESSON PRACTICE

Add using the Rule of Four. The first one has been done for you.

1.

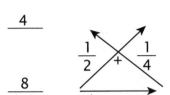

$$\frac{4}{8} + \frac{2}{8} = \frac{6}{8}$$

2.

3.

Subtract using the Rule of Four.

4.

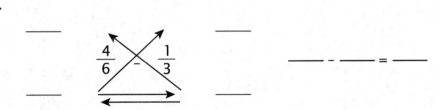

5.

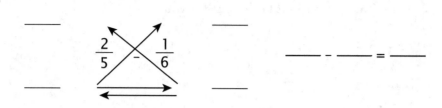

6.

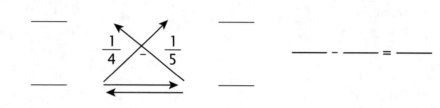

LESSON PRACTICE

Add using the Rule of Four. These are similar to the problems on 6A, except that the arrows have not been drawn to show the steps. The first one has been done for you.

1.

$\underline{\quad 8 \quad}$ $\dfrac{2}{5}$ + $\dfrac{1}{4}$ $\underline{\quad 5 \quad}$ $\dfrac{8}{20} + \dfrac{5}{20} = \dfrac{13}{20}$

$\underline{\quad 20 \quad}$ $\underline{\quad 20 \quad}$

2.

$\underline{\quad\quad}$ $\dfrac{1}{4}$ + $\dfrac{3}{5}$ $\underline{\quad\quad}$ $\underline{\quad\quad} + \underline{\quad\quad} = \underline{\quad\quad}$

$\underline{\quad\quad}$ $\underline{\quad\quad}$

3.

$\underline{\quad\quad}$ $\dfrac{1}{6}$ + $\dfrac{1}{3}$ $\underline{\quad\quad}$ $\underline{\quad\quad} + \underline{\quad\quad} = \underline{\quad\quad}$

$\underline{\quad\quad}$ $\underline{\quad\quad}$

4.

$\underline{\quad\quad}$ $\dfrac{2}{3}$ + $\dfrac{1}{4}$ $\underline{\quad\quad}$ $\underline{\quad\quad} + \underline{\quad\quad} = \underline{\quad\quad}$

$\underline{\quad\quad}$ $\underline{\quad\quad}$

Subtract using the Rule of Four.

5.

$$\frac{4}{6} - \frac{1}{5}$$

$$\underline{\qquad} - \underline{\qquad} = \underline{\qquad}$$

6.

$$\frac{1}{2} - \frac{1}{4}$$

$$\underline{\qquad} - \underline{\qquad} = \underline{\qquad}$$

7.

$$\frac{4}{5} - \frac{2}{3}$$

$$\underline{\qquad} - \underline{\qquad} = \underline{\qquad}$$

8.

$$\frac{2}{4} - \frac{1}{3}$$

$$\underline{\qquad} - \underline{\qquad} = \underline{\qquad}$$

9. Sarah was so hungry that she ate $\frac{3}{4}$ of the casserole for dinner and $\frac{1}{6}$ for a bedtime snack. What part of the casserole did she eat?

10. Yesterday it rained $\frac{3}{4}$ of an inch. Today we got only $\frac{1}{2}$ of an inch of rain. How much less rain did we get today than yesterday?

LESSON PRACTICE

Add using the Rule of Four. Using this method, you can solve problems that cannot be built with the overlays.

1.

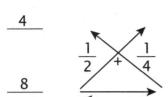

$$\frac{4}{8} \qquad \frac{2}{8}$$

$$\underline{\quad} + \underline{\quad} = \underline{\quad}$$

2.

$$\underline{\quad} \quad \frac{1}{7} + \frac{2}{5} \quad \underline{\quad} \qquad \underline{\quad} + \underline{\quad} = \underline{\quad}$$

3.

$$\underline{\quad} \quad \frac{3}{8} + \frac{1}{2} \quad \underline{\quad} \qquad \underline{\quad} + \underline{\quad} = \underline{\quad}$$

4.

$$\underline{\quad} \quad \frac{1}{6} + \frac{3}{4} \quad \underline{\quad} \qquad \underline{\quad} + \underline{\quad} = \underline{\quad}$$

Subtract using the Rule of Four.

5.

$$\frac{5}{6} - \frac{3}{5}$$

_____ - _____ = _____

6.

$$\frac{4}{5} - \frac{1}{4}$$

_____ - _____ = _____

7.

$$\frac{1}{3} - \frac{1}{5}$$

_____ - _____ = _____

8.

$$\frac{7}{9} - \frac{1}{2}$$

_____ - _____ = _____

9. One half of the books in Nena's library were biographies, and one fourth of them were poetry. What part of her books was either biography or poetry?

10. Cassie filled $\frac{2}{5}$ of her bucket with blueberries. Andrew filled $\frac{1}{3}$ of his bucket with berries. How many more blueberries does Cassie have?

Add or subtract using the Rule of Four.

1. $\dfrac{3}{4} + \dfrac{1}{5} =$ ——

2. $\dfrac{1}{4} + \dfrac{2}{6} =$ ——

3. $\dfrac{2}{3} + \dfrac{1}{6} =$ ——

4. $\dfrac{1}{4} - \dfrac{1}{6} =$ ——

5. $\dfrac{2}{6} - \dfrac{1}{5} =$ ——

6. $\dfrac{4}{9} - \dfrac{1}{8} =$ ——

Fill in the missing numbers in the numerators or denominators to make equivalent fractions.

7. $\dfrac{1}{5} = \dfrac{}{10} = \dfrac{}{15} = \dfrac{4}{}$

8. $\dfrac{4}{5} = \dfrac{}{} = \dfrac{}{} = \dfrac{}{20}$

Solve.

9. $\dfrac{1}{3}$ of 12 = ____

10. $\dfrac{7}{8}$ of 16 = ____

11. $\dfrac{4}{9}$ of 81 = ____

QUICK REVIEW

Rounding and estimation may be used to see if your answer to a multiplication problem is sensible. Remember to regroup in the proper column. Study the example below.

Example 1

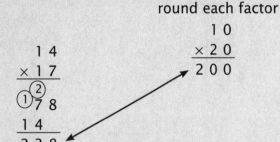

round each factor

$$\begin{array}{r} 1\,0 \\ \times\,2\,0 \\ \hline 2\,0\,0 \end{array}$$

Multiply 1×2 and then write the number of zeros in the two factors.

Compare the estimated answer with the actual answer.

The closer the rounded numbers are to the actual numbers, the closer the estimate will be to the actual answer.

Estimate and then multiply to find the exact answer.

12. $\begin{array}{r} 2\,3 \\ \times\,3\,6 \end{array} \rightarrow$

13. $\begin{array}{r} 7\,8 \\ \times\,3\,4 \end{array} \rightarrow$

14. $\begin{array}{r} 6\,5 \\ \times\,1\,5 \end{array} \rightarrow$

15. Brooke picked 15 tomatoes from each of her 24 tomato plants. How many tomatoes did she pick?

16. Kelsey has $\frac{2}{3}$ of the money she needs. If the total amount needed is $9, how much money does Kelsey have? How much does she still need?

17. One fifth of King Henry's troops were immediately driven away from their siege of the castle. Soon one sixth more of his troops retreated. What part of the troops has been driven away from the battle?

18. This week we had $\frac{1}{2}$ of an inch of rain on Tuesday and $\frac{2}{5}$ of an inch on Thursday. How much rain has fallen this week?

SYSTEMATIC REVIEW

Add or subtract using the Rule of Four.

1. $\dfrac{1}{5} + \dfrac{3}{6} = $ ——

2. $\dfrac{1}{7} + \dfrac{2}{3} = $ ——

3. $\dfrac{1}{3} + \dfrac{1}{5} = $ ——

4. $\dfrac{5}{6} - \dfrac{2}{3} = $ ——

5. $\dfrac{1}{3} - \dfrac{1}{4} = $ ——

6. $\dfrac{1}{2} - \dfrac{1}{9} = $ ——

Fill in the missing numbers in the numerators or denominators to make equivalent fractions.

7. $\dfrac{3}{7} = $ —— $= \dfrac{}{21} = \dfrac{12}{}$

8. $\dfrac{9}{11} = $ —— $= $ —— $= \dfrac{}{44}$

Solve.

9. $\dfrac{1}{7}$ of 35 = ____

10. $\dfrac{4}{5}$ of 50 = ____

11. $\dfrac{3}{4}$ of 36 = ____

Estimate and then multiply to find the exact answer.

12. $\begin{array}{r} 45 \rightarrow \\ \times 24 \end{array}$ 13. $\begin{array}{r} 67 \rightarrow \\ \times 18 \end{array}$

14. $\begin{array}{r} 32 \rightarrow \\ \times 39 \end{array}$

15. Rachel earned $33 dollars a day for 15 days. How much did she earn?

16. Shawn needs 80 posts for the fence he is putting up. He has only $\frac{9}{10}$ of the amount he needs. How many posts does Shawn have?

17. Three sixths of the pie had been eaten. Mom gave one sixth of a whole pie to Jeremy. What part of the the pie is gone now? What part of the pie is still left to be eaten?

18. Clyde grew $\frac{1}{12}$ of a foot last month and $\frac{1}{6}$ of a foot this month. How much more did he grow this month than last month?

19. Round 531 to the nearest hundred.

20. One side of a square is six yards long. What is the perimeter of the square?

SYSTEMATIC REVIEW

Add or subtract using the Rule of Four.

1. $\dfrac{4}{6} + \dfrac{1}{5} = $ ——

2. $\dfrac{2}{4} + \dfrac{1}{3} = $ ——

3. $\dfrac{2}{7} + \dfrac{3}{11} = $ ——

4. $\dfrac{1}{2} - \dfrac{1}{5} = $ ——

5. $\dfrac{2}{3} - \dfrac{1}{4} = $ ——

6. $\dfrac{2}{8} - \dfrac{2}{9} = $ ——

Fill in the missing numbers in the numerators or denominators to make equivalent fractions.

7. $\dfrac{5}{6} = \dfrac{\quad}{\quad} = \dfrac{\quad}{18} = \dfrac{20}{\quad}$

8. $\dfrac{1}{10} = \dfrac{\quad}{\quad} = \dfrac{\quad}{\quad} = \dfrac{\quad}{40}$

Solve.

9. $\dfrac{3}{7}$ of 28 = ____

10. $\dfrac{1}{6}$ of 54 = ____

11. $\dfrac{4}{8}$ of 8 = ____

Estimate and then multiply to find the exact answer.

12.　　7 3 →
　　　× 8 9

13.　　2 6 →
　　　× 9 1

14.　　4 7 →
　　　× 1 1

15. Debbie bought 18 dozen eggs to make baked goods for a wedding reception. How many eggs does she have in all?

16. Bria drove 312 miles every day for three days. How far did she drive?

17. Gretchen used $\frac{4}{9}$ of a loaf of bread to make French toast and $\frac{3}{6}$ of a loaf to make sandwiches. What part of a loaf did she use?

18. Aaron practiced the piano for $\frac{5}{6}$ of an hour today. For how many minutes did he practice?

19. Round 250 to the nearest hundred.

20. Ryan is making a rectangular picture frame for a picture that is 13 inches wide and 18 inches high. He can buy the wood in pieces that are 24 inches long. Will two pieces be long enough to make his frame? (First find the perimeter of his picture. Then find out if 2 × 24 will be long enough.)

Any object or group of objects that can be divided into equal parts may be used to illustrate a fraction. These drawings can help you show $\frac{1}{2}$ in different ways.

1. The square is shaded to show $\frac{1}{2}$. Shade or circle the other drawings to show $\frac{1}{2}$ of the whole number or the whole shape.

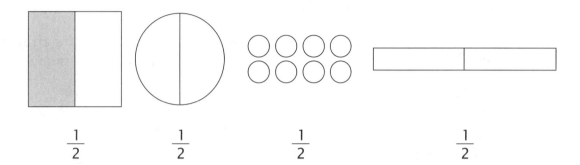

2. Shade the drawings to match problem 1. Then draw lines to show that $\frac{2}{4}$ is the same as $\frac{1}{2}$.

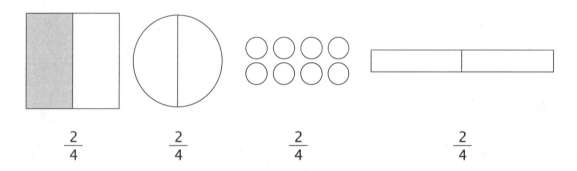

3. Shade or circle the drawings to show how many eighths are the same as $\frac{1}{4}$. Fill in the numerator of the second fraction under each drawing to tell how many eighths are shown.

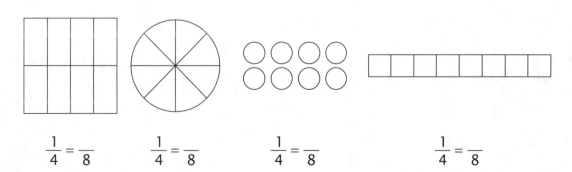

Application and Enrichment 3G showed how parentheses can tell you what part of a math problem to do first. Use the parentheses to help you solve these problems. Numbers 4 and 5 have been done for you.

4. $4 \times (9 + 1) = 4 \times (10) = 40$

5. Write a word problem that fits the equation in #4.

    Example: "Justin found nine spiders this morning and one in the afternoon. Josiah found four times as many spiders as Justin found. How many spiders did Josiah find?"

6. $3 \times (6 - 4) =$

7. Write a word problem that fits the equation in #6.

8. $5 \times (2 + 3 + 1) =$

9. Write a word problem that fits the equation in #8.

10. $(1 + 2) \times (4 + 2) =$

11. Write a word problem that fits the equation in #10.

Build the problems. Use the Rule of Four. Then compare the fractions. Write >, <, or = in the ovals. The first one has been done for you.

1.

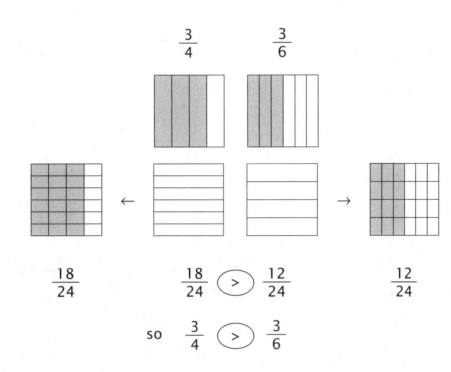

2.

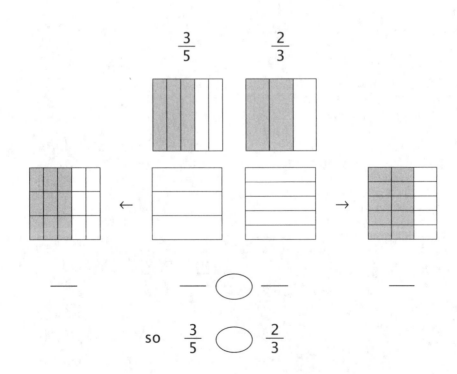

Build the problems. Use the Rule of Four. Then compare the fractions. Write >, <, or = in the ovals. The first one has been done for you.

3.

$\dfrac{4}{}$ 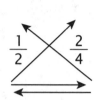 $\dfrac{4}{}$

$\dfrac{1}{2}$ $\dfrac{2}{4}$

$\dfrac{8}{}$ $\dfrac{8}{}$

$\dfrac{4}{8}$ $\boxed{=}$ $\dfrac{4}{8}$ so $\dfrac{1}{2}$ $\boxed{=}$ $\dfrac{2}{4}$

4.

$\dfrac{3}{5}$ $\dfrac{4}{6}$

— ◯ — so $\dfrac{3}{5}$ ◯ $\dfrac{4}{6}$

5.

$\dfrac{2}{3}$ $\dfrac{3}{4}$

— ◯ — so $\dfrac{2}{3}$ ◯ $\dfrac{3}{4}$

6.

$\dfrac{2}{5}$ $\dfrac{1}{3}$

— ◯ — so $\dfrac{2}{5}$ ◯ $\dfrac{1}{3}$

LESSON PRACTICE

Build the problems. Compare using the Rule of Four. Write >, <, or = in the ovals and the correct words in the blanks. The first one has been done for you.

1.

$$\frac{3}{} \qquad \frac{1}{2} \quad \frac{1}{3} \qquad \frac{2}{}$$

$$\underline{6} \qquad \qquad \qquad \underline{6}$$

$$\frac{3}{6} \;\boxed{>}\; \frac{2}{6} \quad \text{so} \quad \frac{1}{2} \;\boxed{>}\; \frac{1}{3}$$

One half is __greater than__ one third.

2.

$$\underline{} \qquad \frac{2}{3} \quad \frac{5}{6} \qquad \underline{}$$

$$\underline{} \qquad \qquad \qquad \underline{}$$

$$\text{---}\bigcirc\text{---} \quad \text{so} \quad \frac{2}{3} \;\bigcirc\; \frac{5}{6}$$

Two thirds is _____ five sixths.

3.

$$\underline{} \qquad \frac{2}{3} \quad \frac{3}{6} \qquad \underline{}$$

$$\underline{} \qquad \qquad \qquad \underline{}$$

$$\text{---}\bigcirc\text{---} \quad \text{so} \quad \frac{2}{3} \;\bigcirc\; \frac{3}{6}$$

Two thirds is _____ three sixths.

4.

$$\underline{} \qquad \frac{1}{2} \quad \frac{2}{5} \qquad \underline{}$$

$$\underline{} \qquad \qquad \qquad \underline{}$$

$$\text{---}\bigcirc\text{---} \quad \text{so} \quad \frac{1}{2} \;\bigcirc\; \frac{2}{5}$$

One half is _____ two fifths.

5.

$$\frac{1}{3} \qquad \frac{2}{6}$$

so $\frac{1}{3}$ ◯ $\frac{2}{6}$

One third is _____ two sixths.

6.

$$\frac{2}{4} \qquad \frac{1}{5}$$

so $\frac{2}{4}$ ◯ $\frac{1}{5}$

Two fourths is _____ one fifth.

7. One half of the students voted for Trisha as class president, while two fifths of them voted for Tom. Which person ended up with more votes?

8. Mike ran $\frac{2}{6}$ of a mile, and Donald ran $\frac{2}{3}$ of a mile. Which person ran the greater distance?

Not all of these problems can be built with the fraction overlays. Compare the fractions using the Rule of Four. Write >, <, or = in the ovals and the correct words in the blanks.

1.

____

$\frac{4}{5}$ $\frac{4}{6}$ ____

____

— ◯ — so $\frac{4}{5}$ ◯ $\frac{4}{6}$

Four fifths is _____ four sixths.

2.

____

$\frac{4}{6}$ $\frac{2}{2}$

____   ____

— ◯ — so $\frac{4}{6}$ ◯ $\frac{2}{2}$

Four sixths is _____ two halves.

3.

____   ____

$\frac{3}{8}$ $\frac{4}{7}$

____   ____

— ◯ — so $\frac{3}{8}$ ◯ $\frac{4}{7}$

Three eighths is_____ four sevenths.

4.

____   ____

$\frac{2}{9}$ $\frac{1}{3}$

____   ____

— ◯ — so $\frac{2}{9}$ ◯ $\frac{1}{3}$

Two ninths is _____ one third.

Use the Rule of Four to compare the fractions. Write >, <, or = in the ovals.

5. $\frac{3}{4}$ $\bigcirc$ $\frac{5}{6}$

6. $\frac{3}{6}$ $\bigcirc$ $\frac{2}{4}$

7. $\frac{1}{2}$ $\bigcirc$ $\frac{3}{10}$

8. $\frac{4}{5}$ $\bigcirc$ $\frac{6}{7}$

9. Shirley ate one fourth of a pizza, and Andrea ate one sixth of a pizza. Which girl ate more pizza?

10. Jeremiah had $\frac{3}{5}$ of an acre of land on the east side of the road and $\frac{7}{12}$ of an acre on the west side. Which was the larger piece of land?

Use the Rule of Four to compare the fractions. Write >, <, or = in the ovals.

1. $\frac{1}{3}$ $\bigcirc$ $\frac{3}{6}$ 2. $\frac{5}{8}$ $\bigcirc$ $\frac{1}{2}$

3. $\frac{3}{12}$ $\bigcirc$ $\frac{1}{4}$

Add or subtract.

4. $\frac{2}{4} + \frac{1}{6} =$ ——   5. $\frac{6}{10} - \frac{3}{8} =$ ——

6. $\frac{2}{9} + \frac{5}{7} =$ ——

Fill in the missing numbers in the numerators or denominators to make equivalent fractions.

7. $\frac{6}{8} = \frac{}{16} = \frac{}{24} = \frac{24}{}$

Solve.

8. $\frac{1}{2}$ of 6 = ____   9. $\frac{3}{6}$ of 42 = ____

10. $\frac{3}{8}$ of 24 = ____

QUICK REVIEW

When the final remainder of a division problem doesn't divide evenly, you may write it as a fraction by writing the remainder over the divisor. Add the resulting fraction to your answer to make a mixed number. Look carefully at the example that has been done for you.

Divide fully. The first one has been done for you.

11.
$$4 \overline{\smash{)}26} = 6\frac{2}{4}$$
$$\underline{24}$$
$$2$$

12. $5 \overline{\smash{)}23}$

13. $7 \overline{\smash{)}59}$

14. Alaina had 17 yards of fabric. She divided it into four equal parts to make curtains. How many yards of fabric does she have for each curtain? Include a fraction in your answer if you are unable to divide evenly.

15. Brad has completed $\frac{2}{7}$ of the chores, and Penny has done $\frac{5}{8}$ of them. Which person has completed the most chores? What part of the chores remains to be finished?

16. If Brad and Penny had a total of 56 chores to do, how many actual chores remain to be done? (See #15.)

17. One fourth of a cup of brown sugar is needed for one recipe, and one third of a cup is needed for another. How much brown sugar is needed in all?

18. During the first storm, $\frac{1}{3}$ of an inch of rain fell. The second storm gave us $\frac{7}{8}$ of an inch of rain. How much more rain fell during the second storm than during the first?

Use the Rule of Four to compare the fractions. Write >, <, or = in the ovals.

1. $\dfrac{3}{5}$ ◯ $\dfrac{1}{3}$

2. $\dfrac{2}{3}$ ◯ $\dfrac{1}{6}$

3. $\dfrac{9}{10}$ ◯ $\dfrac{7}{12}$

Add or subtract.

4. $\dfrac{1}{2} + \dfrac{2}{5} = $ ——

5. $\dfrac{2}{4} - \dfrac{1}{3} = $ ——

6. $\dfrac{3}{8} + \dfrac{3}{5} = $ ——

Fill in the missing numbers in the numerators or denominators to make equivalent fractions.

7. $\dfrac{1}{10} = \dfrac{\quad}{\quad} = \dfrac{\quad}{\quad} = \dfrac{4}{\quad}$

Solve.

8. $\dfrac{7}{8}$ of 32 = ____

9. $\dfrac{2}{7}$ of 21 = ____

10. $\dfrac{3}{4}$ of 20 = ____

Divide. Include a fraction in the answer if you are unable to divide evenly.

11. $6\overline{)32}$

12. $8\overline{)19}$

13. $5\overline{)48}$

Estimate and then multiply to find the exact answer.

14.
$$\begin{array}{r} 21 \\ \times\,16 \\ \hline \end{array}$$

15.
$$\begin{array}{r} 34 \\ \times\,29 \\ \hline \end{array}$$

16.
$$\begin{array}{r} 75 \\ \times\,12 \\ \hline \end{array}$$

17. One sixth of the cars that Valerie saw on her vacation were red, and one seventh of them were blue. What part of the cars that she saw was either red or blue?

18. Luke's team won $\frac{4}{7}$ of the games they played this season. If they played 28 games, how many did they win?

19. Evan's rectangular lawn measures 8 yards by 10 yards. He planted a hedge along $\frac{1}{4}$ of the perimeter. How long was his hedge?

20. Last week's storm gave us one half foot of snow. This week we had a storm that dropped three eighths of a foot. Write a comparison of the two amounts of snow from the storms using >, <, or =.

Use the Rule of Four to compare the fractions. Write >, <, or = in the ovals.

1. $\dfrac{5}{10}$ ◯ $\dfrac{6}{12}$

2. $\dfrac{2}{7}$ ◯ $\dfrac{3}{5}$

3. $\dfrac{1}{2}$ ◯ $\dfrac{2}{3}$

Add or subtract.

4. $\dfrac{2}{3} + \dfrac{1}{5} =$ ____

5. $\dfrac{4}{6} - \dfrac{1}{4} =$ ____

6. $\dfrac{5}{6} + \dfrac{1}{9} =$ ____

Fill in the missing numbers in the numerators or denominators to make equivalent fractions.

7. $\dfrac{3}{4} =$ ____ = ____ = ____

Solve.

8. $\dfrac{3}{5}$ of 10 = ____

9. $\dfrac{1}{4}$ of 12 = ____

10. $\dfrac{4}{6}$ of 24 = ____

Divide. Include a fraction in the answer if you are unable to divide evenly.

11. $3\overline{)13}$ 12. $4\overline{)39}$

13. $9\overline{)58}$

Estimate and then multiply to find the exact answer.

14. $\begin{array}{r} 64 \\ \times\,51 \\ \hline \end{array}$ 15. $\begin{array}{r} 45 \\ \times\,19 \\ \hline \end{array}$

16. $\begin{array}{r} 82 \\ \times\,37 \\ \hline \end{array}$

17. What is the perimeter of a triangle with sides that measure 8 feet, 9 feet, and 10 feet?

18. Kiley answered $\frac{5}{6}$ of the test questions correctly, while Casey answered $\frac{4}{5}$ of the questions on the same test correctly. Write a comparison to show who had more correct answers.

19. If there were 30 questions on the test in #18, how many questions did each girl answer correctly? Write another comparison using the actual numbers. Does it agree with the comparison you wrote for #18?

20. Faith has finished $\frac{5}{8}$ of her chores, and Colleen has finished $\frac{3}{4}$ of her chores. Write a comparison to show the progress of the two girls in finishing their chores.

Decide whether you should add or subtract to solve each problem. Follow the directions to put the correct letters in the blanks.

$$\frac{}{1} \ \frac{}{2} \ \frac{}{3} \ \frac{}{4} \qquad \frac{}{5} \ \frac{}{6} \ \frac{}{7}$$

$$\frac{}{8} \ \frac{}{9} \ \frac{}{10} \ \frac{}{11} \ \frac{}{12} \ \frac{}{13} \ \frac{}{14}.$$

1. A fraction of the job was done before lunch, and another fraction was finished after lunch. What part of the job has been finished?

    For addition, put R in blanks 1 and 7.

    For subtraction, put S in blank 1.

2. Jim ate one part of the pizza, and Bob ate another part. What part of the whole pizza has been eaten?

    For addition, put E in blanks 2 and 9.

    For subtraction, put T in blank 2.

3. Ava painted a fraction of the fence yesterday. When she finished work today, all of the fence was painted. What part of the fence was painted today?

    For addition, put O in blanks 3 and 6.

    For subtraction, put A in blanks 3 and 10.

4. One part of the books was biographies, and another part was poetry. What part of the books was either biography or poetry?

    For addition, put D in blank 4.

    For subtraction, put P in blank 4.

5. Yesterday it rained a fraction of an inch. Today it rained another fraction of an inch. How much more rain did we get today than we got yesterday?

    For addition, put R in blank 7 and 8.

    For subtraction, put F in blank 5.

6. Michael grew a fraction of a foot taller last year. Gabriel grew a different fraction of a foot taller the same year. What is the difference in the amount they grew?

    For addition, put F in blank 5.

    For subtraction, put O in blank 6.

7. Amy needs a fraction of a cup of honey for bread and a fraction of a cup for cookies. How much honey does she need altogether?

    For addition, put M in blank 8.

    For subtraction, put E in blanks 9 and 12.

8. Micah walked a fraction of a mile, and Brandon walked a greater fraction of a mile. How much farther did Brandon walk?

    For addition, put V in blank 10.

    For subtraction, put N in blanks 11 and 13.

9. Sammy ate a fraction of the cookies, and Tom ate another fraction of the cookies. What fraction tells the total part of the cookies that were eaten?

    For addition, put I in blank 12.

    For subtraction, put I in blank 11.

10. Kim did part of the job on Monday and another part on Tuesday. What part of the job has she completed?

    For addition, put G in blank 14.

    For subtraction, put W in blank 13 and leave blank 14 empty.

Use Method 1 to add the three fractions. The first one has been done for you.

1. $\dfrac{1}{3} + \dfrac{3}{4} + \dfrac{1}{2} = \dfrac{38}{24}$

2. $\dfrac{1}{5} + \dfrac{1}{6} + \dfrac{1}{2} = $ ____

$\overset{4}{\underset{12}{}}\dfrac{1}{3} + \dfrac{3}{4}\overset{9}{\underset{12}{}} = \dfrac{13}{12}$

$\overset{26}{\underset{24}{}}\dfrac{13}{12} + \dfrac{1}{2}\overset{12}{\underset{24}{}} = \dfrac{38}{24}$

3. $\dfrac{1}{8} + \dfrac{1}{3} + \dfrac{1}{6} = $ ____

4. $\dfrac{1}{4} + \dfrac{5}{6} + \dfrac{1}{3} = $ ____

Use Method 2 or Method 3 to add the three fractions. For now, you may keep the fraction with the greater number on top or simplify it by dividing. The first two have been done for you.

5. $\dfrac{1}{4} + \dfrac{2}{3} + \dfrac{1}{5} = \dfrac{67}{60}$

$\dfrac{1 \times 3 \times 5}{4 \times 3 \times 5} + \dfrac{2 \times 4 \times 5}{3 \times 4 \times 5} + \dfrac{1 \times 4 \times 3}{5 \times 4 \times 3} = \dfrac{15}{60} + \dfrac{40}{60} + \dfrac{12}{60} = \dfrac{67}{60}$

6. $\dfrac{7}{10} + \dfrac{1}{2} + \dfrac{4}{5} = \dfrac{20}{10}$ or 2

$\dfrac{7}{10} + \dfrac{1 \times 5}{2 \times 5} + \dfrac{4 \times 2}{5 \times 2} = \dfrac{7}{10} + \dfrac{5}{10} + \dfrac{8}{10} = \dfrac{20}{10}$ or 2

7. $\dfrac{3}{7} + \dfrac{1}{6} + \dfrac{3}{10} =$ ———

8. $\dfrac{4}{5} + \dfrac{3}{10} + \dfrac{1}{2} =$ ———

Use the method you prefer to solve the word problems.

9. A plumber needs the following lengths of pipe: $\frac{5}{8}$ ft, $\frac{7}{8}$ ft, and $\frac{1}{2}$ ft. What is the total length of pipe he needs?

10. Haley's car used $\frac{1}{8}$ of a tank of gas for the first part of her trip, $\frac{1}{5}$ of a tank for the second part, and $\frac{1}{4}$ of a tank for the third part. What part of a tank of gas was used for the whole trip?

LESSON PRACTICE

Use Method 1 to add the three fractions.

1. $\dfrac{1}{2} + \dfrac{2}{5} + \dfrac{5}{6} =$ ___

2. $\dfrac{2}{7} + \dfrac{1}{5} + \dfrac{2}{3} =$ ___

3. $\dfrac{3}{8} + \dfrac{1}{2} + \dfrac{2}{5} =$ ___

4. $\dfrac{1}{7} + \dfrac{2}{9} + \dfrac{1}{3} =$ ___

Use Method 2 or Method 3 to add the three fractions.

5. $\dfrac{5}{7} + \dfrac{1}{2} + \dfrac{1}{3} =$ ___

6. $\dfrac{1}{2} + \dfrac{3}{4} + \dfrac{5}{8} =$ ___

7. $\dfrac{2}{5} + \dfrac{1}{4} + \dfrac{1}{6} =$ ___

8. $\dfrac{4}{9} + \dfrac{1}{3} + \dfrac{2}{9} =$ ___

Use the method you prefer to solve the word problems.

9. Alex ran for $\frac{1}{10}$ of a mile, jogged for $\frac{2}{5}$ of a mile, and walked for $\frac{1}{2}$ of a mile. How far did he travel?

10. Mom passed out orange sections. Luke had $\frac{5}{6}$ of an orange, Seth had $\frac{1}{2}$, and Karen had $\frac{2}{3}$. How many oranges were eaten? Divide the numerator by the denominator to find the final answer.

Use the method you prefer to add the fractions.

1. $\dfrac{3}{5} + \dfrac{1}{2} + \dfrac{1}{3} =$ ——

2. $\dfrac{5}{6} + \dfrac{1}{3} + \dfrac{2}{5} =$ ——

3. $\dfrac{1}{3} + \dfrac{3}{4} + \dfrac{5}{6} =$ ——

4. $\dfrac{2}{6} + \dfrac{1}{3} + \dfrac{1}{2} =$ ——

5. $\dfrac{1}{5} + \dfrac{1}{5} + \dfrac{3}{10} =$ ——

6. $\dfrac{2}{5} + \dfrac{1}{4} + \dfrac{1}{6} =$ ——

7. $\dfrac{6}{7} + \dfrac{3}{14} + \dfrac{1}{2} =$ ——

8. $\dfrac{7}{8} + \dfrac{1}{4} + \dfrac{1}{3} =$ ——

9. One fourth of the job has been assigned to David, one eighth of it has been assigned to Daniel, and one fourth of it has been assigned to Douglas. What part of the job has been assigned?

10. Meredith spent $\frac{1}{2}$ of her birthday money on books, $\frac{1}{4}$ on clothes, and $\frac{1}{8}$ on a treat for her brothers. What part of her money has she spent so far?

Use the method you prefer to add the fractions.

1. $\dfrac{2}{5} + \dfrac{1}{4} + \dfrac{1}{2} =$ ____

2. $\dfrac{9}{10} + \dfrac{1}{2} + \dfrac{3}{4} =$ ____

3. $\dfrac{1}{6} + \dfrac{2}{3} + \dfrac{1}{2} =$ ____

Use the Rule of Four to compare the fractions. Write >, <, or = in the ovals.

4. $\dfrac{1}{4} \bigcirc \dfrac{1}{6}$

5. $\dfrac{3}{5} \bigcirc \dfrac{1}{3}$

6. $\dfrac{3}{6} \bigcirc \dfrac{2}{4}$

Fill in the missing numbers in the numerators or denominators to make equivalent fractions.

7. $\dfrac{2}{3} =$ ____ $=$ ____ $= \dfrac{}{12}$

8. ____ $= \dfrac{2}{12} =$ ____ $= \dfrac{4}{}$

Multiply. Use estimation if you wish.

9. $\begin{array}{r} 7\,3 \\ \times\,6\,2 \\ \hline \end{array}$

10. $\begin{array}{r} 5\,4 \\ \times\,2\,8 \\ \hline \end{array}$

11. $\begin{array}{r} 9\,1 \\ \times\,4\,9 \\ \hline \end{array}$

QUICK REVIEW

Here are some division problems to review. Study the example.

Divide. Include a fraction in the answer you are unable to divide evenly. The first one has been done for you.

12.
$$5\overline{)264} = 52\tfrac{4}{5}$$
```
        5 2  4/5
     5 | 2 6 4
         2 5 0
           1 4
           1 0
              4
```

13. $6\overline{)379}$

14. $2\overline{)503}$

15. One half of the bird's nest was made of sticks, one eighth was made of grass, and one sixteenth was made of hair. Was that the entire nest, or must some other material have been used as well?

16. The local chorus divided themselves into groups of four to sing Christmas carols. If there were 35 people in the chorus, how many groups were they able to form? How many people were left over?

   (Do not use a fraction when writing your answer in this case!)

17. Ian enjoyed $\frac{3}{4}$ of the books his sister read to him. If she read eight books, how many did he enjoy?

18. Dad emptied his pocket and gave $\frac{1}{6}$ of the coins he found to Ben and $\frac{2}{7}$ of them to Clara. Which person received more coins?

Use the method you prefer to add the fractions.

1. $\dfrac{5}{8} + \dfrac{1}{4} + \dfrac{1}{2} = $ ——

2. $\dfrac{2}{10} + \dfrac{1}{2} + \dfrac{5}{6} = $ ——

3. $\dfrac{1}{2} + \dfrac{1}{3} + \dfrac{3}{6} = $ ——

Use the Rule of Four to compare the fractions. Write >, <, or = in the ovals.

4. $\dfrac{2}{3} \bigcirc \dfrac{2}{6}$

5. $\dfrac{5}{8} \bigcirc \dfrac{1}{4}$

6. $\dfrac{2}{5} \bigcirc \dfrac{3}{7}$

Fill in the missing numbers in the numerators or denominators to make equivalent fractions.

7. $\dfrac{1}{8} = $ —— $ = $ —— $ = \dfrac{}{32}$

8. —— $ = \dfrac{2}{4} = $ —— $ = \dfrac{4}{}$

Multiply. Use estimation if you wish.

9. $\begin{array}{r} 3\,5 \\ \times\,2\,2 \\ \hline \end{array}$

10. $\begin{array}{r} 4\,7 \\ \times\,8\,4 \\ \hline \end{array}$

11. $\begin{array}{r} 6\,3 \\ \times\,1\,9 \\ \hline \end{array}$

Divide.

12. $3\overline{)198}$

13. $8\overline{)809}$

14. $4\overline{)472}$

15. Fred ate $\frac{1}{2}$ of a pizza for breakfast, $\frac{1}{3}$ of a pizza for lunch, and $\frac{1}{6}$ of a pizza for dinner. How much pizza did he eat today?

16. Andrew ate one half of a pie. Use equivalent fractions to find how many fourths of a pie he ate.

17. Micah bought flashlight batteries for his club's hiking expedition. He needed four batteries for each flashlight, and he bought 45 batteries. How many flashlights would his batteries fill? How many batteries would be left over?

18. On the coldest day of the year, Dianne sold 122 pairs of mittens from her craft shop. How many hands will those mittens keep warm?

19. Anah had 85 dimes. She gave $\frac{2}{5}$ of them to her sister. How many dimes did her sister receive?

20. Jeremy grew $\frac{1}{2}$ of a foot taller last year, and Jeffrey grew $\frac{1}{3}$ of a foot taller. Which boy grew more? Find the difference in their growth and express it as a fraction of a foot.

Use the method you prefer to add the fractions.

1. $\dfrac{7}{10} + \dfrac{3}{4} + \dfrac{1}{3} =$ ——

2. $\dfrac{6}{7} + \dfrac{1}{3} + \dfrac{5}{6} =$ ——

3. $\dfrac{7}{8} + \dfrac{5}{16} + \dfrac{1}{2} =$ ——

Use the Rule of Four to compare the fractions. Write >, <, or = in the ovals.

4. $\dfrac{4}{9}$ $\bigcirc$ $\dfrac{1}{2}$

5. $\dfrac{5}{6}$ $\bigcirc$ $\dfrac{6}{7}$

6. $\dfrac{2}{3}$ $\bigcirc$ $\dfrac{5}{8}$

Fill in the missing numbers in the numerators or denominators to make equivalent fractions.

7. $\dfrac{4}{7} =$ —— $=$ —— $= \dfrac{}{28}$

8. —— $= \dfrac{2}{6} =$ —— $= \dfrac{4}{}$

Multiply. Use estimation if you wish.

9.
```
    3 2
  × 5 5
```

10.
```
    7 6
  × 4 1
```

11.
```
    2 9
  × 1 7
```

Divide.

12. $7\overline{)361}$

13. $9\overline{)734}$

14. $5\overline{)108}$

15. One fifth of Nathan's allowance is budgeted for gifts, one tenth for charity, and one half for saving. What part of his allowance is left for Nathan to spend on himself?

16. What is 85 rounded to the nearest ten?

17. What is 132 rounded to the nearest hundred?

18. A group of soldiers marched around the perimeter of a rectangular area that was five miles long and three miles wide. How many miles did they march?

19. Teresa was sick for $\frac{1}{7}$ of the month of February (28 days). For how many days was she sick?

20. There are 25 chocolates in a box. If they are shared equally among four friends, how many chocolates will each person receive?

In written music, notes have fractional values that tell how long that particular note should be played or sung. The music is divided into measures, and a time signature tells us how many beats are in each measure of that particular song. Here are four commonly-used notes and the values usually given to them.

The time signature tells us how many beats each note is given in a particular song. One common time signature is $\frac{4}{4}$ time. The top number tells us that each measure has four beats. The bottom number tells us that a quarter ($\frac{1}{4}$) note gets one beat. Here is how it works in one measure of a song with $\frac{4}{4}$ time.

The numbers above the notes show the four beats in the measure. The fractions show the names of each note. The two eighth notes are the same as one quarter note because $\frac{1}{8} + \frac{1}{8} = \frac{2}{8}$, which is the same as $\frac{1}{4}$. If you are playing or singing, the last two notes will be played more quickly than the rest of the notes.

Example 1

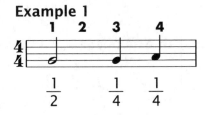

Since $\frac{1}{2}$ is the same as $\frac{1}{4} + \frac{1}{4}$, the half note at the beginning will take twice as long to play or sing. We say that it gets two beats.

The fraction $\frac{4}{4}$ is the same as one. It means that something has been divided into four parts and that we are considering all four of the parts. In music, if the time signature is $\frac{4}{4}$, all the notes in one measure should add up to a fraction that is the same as one. (The first and last measure of a song may not add up to one.)

Looking at Example 1 on the previous page, we add $\frac{1}{2} + \frac{1}{4} + \frac{1}{4}$. Any of the methods from lesson 8 may be used to add the numbers.

Using the Rule of Four and adding two numbers at a time, we get:

$$\frac{1}{2} + \frac{1}{4} = \frac{4}{8} + \frac{2}{8} = \frac{6}{8} \text{ and } \frac{6}{8} + \frac{1}{4} = \frac{24}{32} + \frac{8}{32} = \frac{32}{32}$$

Noticing that the last two fractions have the same denominator, you could also solve this way:

$$\frac{1}{2} + \frac{2}{4} = \frac{4}{8} + \frac{4}{8} = \frac{8}{8}$$

If you remember that $\frac{2}{4}$ is the same as $\frac{1}{2}$, you can add $\frac{1}{2} + \frac{1}{2} = \frac{2}{2}$. Each of the three answers is equal to one because the numerator and the denominator are the same in each case.

Write the correct fraction under each note and add to see if the sum equals one. Look for fractions with the same denominators and add them first. Then add the results.

1.

The sum of the notes is _____ , which is the same as one.

2.

The sum of the notes is _____ , which is the same as one.

3.

The sum of the notes is _____ , which is the same as one.

Build each problem with the overlays. Find a fraction of a fraction. The first one has been done for you.

1. $\dfrac{2}{4}$ of $\dfrac{3}{6}$ = $\dfrac{2 \times 3}{4 \times 6}$ = $\dfrac{6}{24}$

2. $\dfrac{3}{5}$ of $\dfrac{3}{4}$ = _____

3. $\dfrac{1}{3}$ of $\dfrac{1}{2}$ = ___

4. $\dfrac{4}{6}$ of $\dfrac{2}{5}$ = ___

5. $\dfrac{2}{3}$ of $\dfrac{1}{4}$ = ___

6. $\dfrac{1}{5}$ of $\dfrac{5}{6}$ = ___

7. $\dfrac{4}{5}$ of $\dfrac{2}{6}$ = ___

8. $\dfrac{1}{6}$ of $\dfrac{1}{3}$ = ___

9. $\dfrac{1}{2}$ of $\dfrac{2}{6}$ = ___

Build each problem with the overlays. Find a fraction of a fraction.

10. $\dfrac{3}{5} \times \dfrac{2}{5} = $ ——

11. $\dfrac{3}{6} \times \dfrac{2}{3} = $ ——

12. $\dfrac{2}{4} \times \dfrac{1}{5} = $ ——

13. $\dfrac{3}{4} \times \dfrac{4}{6} = $ ——

14. $\dfrac{2}{5} \times \dfrac{1}{2} = $ ——

15. $\dfrac{4}{6} \times \dfrac{1}{3} = $ ——

Always read fraction word problems carefully and think about what is being asked. Multiplication problems involving fractions usually use the word "of," but you should not assume that the word "of" somewhere in a problem automatically indicates multiplication. Read for meaning instead of just looking for key words. All of the problems on lesson practice A, B, and C involve multiplication of fractions.

16. A recipe calls for $\frac{1}{3}$ of a cup of butter. If Matthew is making $\frac{1}{4}$ of the recipe, how much butter should he use?

17. One half of the people at the picnic went home sick, but only one fifth of them were seriously ill. What part of the whole group was seriously ill?

18. Mom left $\frac{1}{3}$ of a pie for Bill to eat, but since he was not very hungry, he ate only $\frac{1}{5}$ of what was there. What part of a whole pie did Bill eat?

LESSON PRACTICE

Find a fraction of a fraction. These can all be built with the overlays.

1. $\dfrac{2}{6}$ of $\dfrac{3}{5}$ = ——

2. $\dfrac{2}{3}$ of $\dfrac{1}{2}$ = ——

3. $\dfrac{3}{8}$ of $\dfrac{2}{4}$ = ——

4. $\dfrac{1}{4}$ of $\dfrac{1}{5}$ = ——

5. $\dfrac{1}{6}$ of $\dfrac{2}{3}$ = ——

6. $\dfrac{1}{3}$ of $\dfrac{4}{5}$ = ——

7. $\dfrac{4}{5}$ of $\dfrac{3}{6}$ = ——

8. $\dfrac{1}{2}$ of $\dfrac{5}{6}$ = ——

9. $\dfrac{3}{4}$ of $\dfrac{3}{4}$ = ——

Find a fraction of a fraction. Some problems cannot be built with the overlays.

10. $\dfrac{2}{5} \times \dfrac{4}{6} = $ ——

11. $\dfrac{1}{7} \times \dfrac{1}{2} = $ ——

12. $\dfrac{3}{5} \times \dfrac{5}{6} = $ ——

13. $\dfrac{3}{10} \times \dfrac{1}{4} = $ ——

14. $\dfrac{1}{4} \times \dfrac{1}{2} = $ ——

15. $\dfrac{7}{8} \times \dfrac{1}{9} = $ ——

16. Three fourths of the class volunteered for the special project, but only $\frac{1}{4}$ of the volunteers were finished with their work and could do the project. What part of the whole class could participate?

17. Joan picked $\frac{2}{3}$ of a bushel of tomatoes and gave $\frac{3}{5}$ of what she had picked to her neighbor. What part of a whole bushel did her neighbor receive?

18. Ryan's team lost $\frac{1}{5}$ of the games they played. Of the games they lost, they did not score in $\frac{1}{5}$ of them. In what part of the total games they played did they fail to score?

LESSON PRACTICE

Find a fraction of a fraction.

1. $\frac{3}{6}$ of $\frac{2}{5}$ = ——

2. $\frac{2}{9}$ of $\frac{2}{3}$ = ——

3. $\frac{2}{6}$ of $\frac{1}{3}$ = ——

4. $\frac{3}{4}$ of $\frac{4}{5}$ = ——

5. $\frac{1}{7}$ of $\frac{3}{4}$ = ——

6. $\frac{1}{6}$ of $\frac{1}{6}$ = ——

7. $\frac{4}{8}$ of $\frac{1}{3}$ = ——

8. $\frac{3}{5}$ of $\frac{3}{4}$ = ——

9. $\frac{4}{6}$ of $\frac{2}{4}$ = ——

Multiply the fractions.

10. $\dfrac{1}{6} \times \dfrac{2}{5} =$ —

11. $\dfrac{5}{9} \times \dfrac{1}{4} =$ —

12. $\dfrac{1}{10} \times \dfrac{1}{10} =$ —

13. $\dfrac{2}{3} \times \dfrac{3}{6} =$ —

14. $\dfrac{3}{5} \times \dfrac{1}{2} =$ —

15. $\dfrac{1}{8} \times \dfrac{3}{5} =$ —

16. It is $\frac{4}{5}$ of a mile around the track. Trevor ran $\frac{2}{4}$ of the way around the track. What part of a mile did he run?

17. Stephanie is making cookies. The recipe calls for $\frac{2}{3}$ of a cup of milk, but Stephanie is making only $\frac{1}{3}$ of the recipe. How much milk is needed?

18. Seven eighths of a pie was left over from Thanksgiving dinner. If Robin ate one seventh of what was there, what part of a whole pie did she eat?

Multiply (fraction of a fraction).

1. $\dfrac{2}{5}$ of $\dfrac{1}{5}$ = ——

2. $\dfrac{5}{6} \times \dfrac{1}{4}$ = ——

3. $\dfrac{5}{9} \times \dfrac{1}{2}$ = ——

Add or subtract.

4. $\dfrac{1}{7} + \dfrac{1}{3}$ = ——

5. $\dfrac{2}{3} - \dfrac{1}{7}$ = ——

6. $\dfrac{1}{2} + \dfrac{2}{5} + \dfrac{7}{10}$ = ——

Use the Rule of Four to make the denominators the same. Then compare the fractions.

7. $\dfrac{4}{6} \bigcirc \dfrac{1}{7}$

8. $\dfrac{2}{3} \bigcirc \dfrac{5}{8}$

9. $\dfrac{1}{9} \bigcirc \dfrac{2}{11}$

Fill in the missing numbers in the numerators or denominators to make equivalent fractions.

10. $\frac{2}{7} = \frac{}{} = \frac{}{} = \frac{}{28}$

11. $\frac{}{} = \frac{2}{18} = \frac{}{} = \frac{4}{}$

QUICK REVIEW

Here is a chance to review multiplying a two-digit number by a three-digit number. The first one has been done for you.

Estimate and then multiply to find the exact answer.

12.

```
      6 2 3          6 0 0
    ×   4 5        ×   5 0
    ---------      ---------
    3 0 0 5        3 0,0 0 0
    2 4 8 2
    ---------
    2 8,0 3 5
```

This is 5 tens × 6 hundreds.

13.
```
    1 7 9
  × 5 7
```

14.
```
    9 0 2
  × 1 1
```

15. Mom is making $\frac{1}{3}$ of a recipe that calls for $\frac{2}{3}$ of a cup of flour. How much flour should she use?

16. There are 365 days in one year. How many days are there in 25 years? (Don't worry about leap years.)

17. Nancy needs $\frac{1}{2}$ cup of honey for bread, $\frac{1}{3}$ cup for cookies, and $\frac{1}{4}$ cup for fruit salad. How much honey does she need altogether?

18. Ivan spotted 36 birds Saturday morning at Hawk Mountain. Three fourths of them were hawks. How many hawks did he see?

SYSTEMATIC REVIEW

Multiply (fraction of a fraction).

1. $\dfrac{1}{2}$ of $\dfrac{2}{3} =$ ——

2. $\dfrac{7}{8} \times \dfrac{2}{5} =$ ——

3. $\dfrac{1}{3} \times \dfrac{3}{5} =$ ——

Add or subtract.

4. $\dfrac{3}{4} + \dfrac{1}{9} =$ ——

5. $\dfrac{2}{3} - \dfrac{2}{5} =$ ——

6. $\dfrac{3}{8} + \dfrac{5}{8} + \dfrac{1}{4} =$ ——

Use the Rule of Four to make the denominators the same. Then compare the fractions.

7. $\dfrac{3}{9} \bigcirc \dfrac{1}{3}$

8. $\dfrac{2}{5} \bigcirc \dfrac{3}{8}$

9. $\dfrac{5}{7} \bigcirc \dfrac{7}{10}$

Estimate and then multiply to find the exact answer.

10.
$$\begin{array}{r} 125 \\ \times\ 51 \\ \hline \end{array}$$

11.
$$\begin{array}{r} 254 \\ \times\ 35 \\ \hline \end{array}$$

12.
$$\begin{array}{r} 563 \\ \times\ 26 \\ \hline \end{array}$$

Divide.

13. $6\overline{)107}$

14. $8\overline{)395}$

15. $2\overline{)459}$

16. Simon has done $\frac{4}{5}$ of his chores. Jill helped him with $\frac{1}{2}$ of what he completed. What part of the total amount of chores did Jill help Simon do?

17. Sally finished $\frac{1}{6}$ of the job, Sarah did $\frac{1}{5}$ of it, and Sue accomplished $\frac{1}{4}$ of the job. What part of the job has been finished?

18. A car went forward 51 inches every time the wheels went around one time. How many inches did the car travel if the wheels went around 250 times?

19. What is the perimeter of a triangle whose sides measure 12 inches, 17 inches, and 28 inches?

20. Christel counted out 235 jelly beans. She gave $\frac{3}{5}$ of them to her mom. How many jelly beans did she give to her mom?

Multiply (fraction of a fraction).

1. $\frac{2}{4}$ of $\frac{2}{5}$ = ——

2. $\frac{1}{3} \times \frac{2}{6}$ = ——

3. $\frac{1}{2} \times \frac{4}{9}$ = ——

Add or subtract.

4. $\frac{2}{3} + \frac{1}{5}$ = ——

5. $\frac{2}{5} - \frac{3}{8}$ = ——

6. $\frac{1}{4} + \frac{3}{5} + \frac{2}{3}$ = ——

Use the Rule of Four to make the denominators the same. Then compare the fractions.

7. $\frac{5}{6}$ ◯ $\frac{4}{8}$

8. $\frac{3}{5}$ ◯ $\frac{4}{9}$

9. $\frac{3}{4}$ ◯ $\frac{6}{8}$

Estimate and then multiply to find the exact answer.

10. 558
    × 62

11. 407
    × 83

12. 349
    × 12

Divide.

13. 7⟌128

14. 3⟌471

15. 5⟌298

16. Three books are stacked in a pile. If each book is $\frac{1}{4}$ of a foot thick, how high is the pile?

17. Caitlyn walked $\frac{4}{5}$ of a mile, while Justin walked $\frac{1}{3}$ of a mile. How much farther did Caitlyn walk?

18. One third of the people in the room have birthdays in the summer. One half of the summer birthdays are in July. If there are 12 people in the room, how many were born in July?

19. What is the perimeter of a triangle whose sides measure 13 inches, 19 inches, and 26 inches?

20. Duncan woke up early on $\frac{1}{2}$ of the days last month. The month had 30 days. On how many days did Duncan wake up early?

Some students are confused because multiplying by a fraction gives an answer that is less than the original amount. It may help to think of multiplying numbers in the following ways:

Two times a number equals an answer that is twice as great as what you started with.

$2 \times 4 = 8$, and 8 is twice as much as 4.
$2 \times \frac{1}{2} = 1$, and 1 is twice as great as $\frac{1}{2}$.

One times a number equals an answer that is the same as what you started with.

$1 \times 4 = 4$, and 4 is the same as 4.
$1 \times \frac{1}{2} = \frac{1}{2}$, and $\frac{1}{2}$ is the same as $\frac{1}{2}$.

One half times a number equals an answer that is one half of what you started with.

$\frac{1}{2} \times 4 = 2$, and 2 is less than 4.
   If 4 items are separated into 2 groups, each group will have 2 pieces.
$\frac{1}{2} \times \frac{1}{2} = \frac{1}{4}$, and $\frac{1}{4}$ is less than $\frac{1}{2}$.
   If $\frac{1}{2}$ is cut in half, each part will be $\frac{1}{4}$ of the whole.

Building problems with the overlays can help. It is also helpful to make drawings. The drawings don't have to be fancy—just make a sketch to show the process. Here are some examples using problems taken from your student book.

Example 1

A recipe calls for $\frac{1}{3}$ of a cup of melted butter. If Matthew is making $\frac{1}{4}$ of the recipe, how much butter should he use?

First draw a cup that is $\frac{1}{3}$ full of butter. Since he is making $\frac{1}{4}$ of the recipe, draw vertical lines to divide the butter into four parts and choose one of the parts. Matthew needs to use $\frac{1}{12}$ of a cup of melted butter. Check it by multiplying: $\frac{1}{4} \times \frac{1}{3} = \frac{1}{12}$.

Example 2

Mom left $\frac{1}{3}$ of a pie for Bill to eat. Bill was not very hungry, so he ate only $\frac{1}{5}$ of what was there. What part of a whole pie did Bill eat?

What Mom left

What Bill ate

$$\frac{1}{5} \times \frac{1}{3} = \frac{1}{15} \text{ of the whole pie}$$

Here are some more problems that you may be able to solve by drawing. Check your answers by multiplying the fractions.

1. A recipe calls for $\frac{1}{3}$ of a cup of milk. Mary is making $\frac{1}{3}$ of the recipe. What part of a cup of milk should she use?

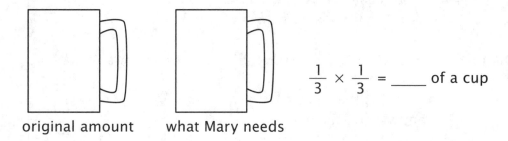

original amount   what Mary needs

$$\frac{1}{3} \times \frac{1}{3} = \rule{2cm}{0.4pt} \text{ of a cup}$$

2. Mary (#1) decided to make two times the recipe instead. Now how much milk does she need?

3. Sam saw 12 birds. Three fourths of them were juncos. How many of the birds were juncos? This problem asks for a fraction of a number, instead of a fraction of one whole. We start by drawing shapes to represent 12 birds. (Lesson 1 tells how to find a fraction of a number.)

$$\frac{3}{4} \times 12 = \rule{2cm}{0.4pt} \text{ juncos}$$

4. Aaron ate pizza every day for three days. Each day he ate $\frac{1}{4}$ of a pizza. How much pizza did he eat in all? Write the whole number 3 as $\frac{3}{1}$.

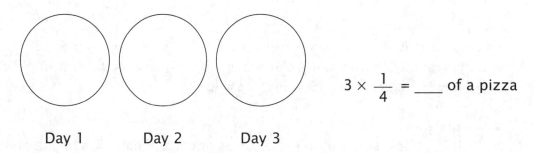

Day 1   Day 2   Day 3

$$3 \times \frac{1}{4} = \rule{2cm}{0.4pt} \text{ of a pizza}$$

The next lesson introduces division of fractions. Be sure you understand the three operations with fractions presented so far. Remember to use drawings whenever you can to make the meaning of a problem clear.

Divide using the Rule of Four. You may write your answer in either form. The first two have been done for you.

1.

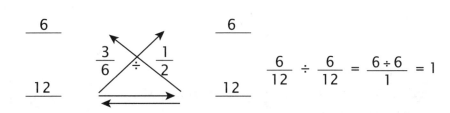

$$\frac{6}{12} \div \frac{6}{12} = \frac{6 \div 6}{1} = 1$$

2.

$$\frac{12}{18} \div \frac{9}{18} = \frac{12 \div 9}{1} = \frac{12}{9} = 1\frac{3}{9}$$

3.

$$\frac{5}{8} \div \frac{2}{3}$$

$$\frac{}{} \div \frac{}{} = \frac{}{1} = \frac{}{}$$

4.

$$\frac{6}{8} \div \frac{1}{2}$$

$$\frac{}{} \div \frac{}{} = \frac{}{1} = \frac{}{}$$

5.

$$\frac{\underline{\quad}}{\underline{\quad}} \qquad \frac{1}{2} \div \frac{1}{4} \qquad \frac{\underline{\quad}}{\underline{\quad}} \qquad \frac{\underline{\quad} \div}{\div \underline{\quad}} = \frac{\underline{\quad}}{1} = \frac{\underline{\quad}}{}$$

Continue to read fraction word problems carefully for meaning. All of the examples on lesson practice A, B, and C involve division of fractions.

6. Sue has $\frac{1}{2}$ pound of nuts in her pantry. How many times can she make a recipe calling for $\frac{1}{8}$ of a pound of nuts?

7. Kara has a ribbon $\frac{2}{3}$ of a yard long. How many pieces $\frac{1}{6}$ of a yard long can she cut from it?

8. Six eighths of the pie is left. Rick wants to give each of his friends one fourth of a whole pie. How many people can he serve?

LESSON PRACTICE

Divide using the Rule of Four.

1.

$$\frac{\underline{}}{\underline{}} \qquad \frac{3}{5} \div \frac{1}{4} \qquad \frac{\underline{}}{\underline{}} \qquad \frac{\underline{} \div}{\underline{} \div} \frac{\underline{}}{\underline{}} = \frac{\underline{}}{1} = \frac{\underline{}}{\underline{}}$$

2.

$$\frac{\underline{}}{\underline{}} \qquad \frac{3}{4} \div \frac{1}{6} \qquad \frac{\underline{}}{\underline{}} \qquad \frac{\underline{} \div}{\underline{} \div} \frac{\underline{}}{\underline{}} = \frac{\underline{}}{1} = \frac{\underline{}}{\underline{}}$$

3.

$$\frac{\underline{}}{\underline{}} \qquad \frac{1}{4} \div \frac{1}{3} \qquad \frac{\underline{}}{\underline{}} \qquad \frac{\underline{} \div}{\underline{} \div} \frac{\underline{}}{\underline{}} = \frac{\underline{}}{1} = \frac{\underline{}}{\underline{}}$$

Notice that the answer to #3 is less than one, so we simply leave the result in fraction form. How many times can we divide $\frac{1}{3}$ of a pie out of $\frac{1}{4}$ of a pie? We can't even get one whole third out of a fourth of a pie.

4.

$$\frac{\underline{}}{\underline{}} \qquad \frac{2}{3} \div \frac{5}{8} \qquad \frac{\underline{}}{\underline{}} \qquad \frac{\underline{} \div}{\underline{} \div} \frac{\underline{}}{\underline{}} = \frac{\underline{}}{1} = \frac{\underline{}}{\underline{}}$$

5.

$$\frac{\underline{}}{\underline{}} \quad \frac{1}{4} \div \frac{4}{5} \quad \frac{\underline{}}{\underline{}} \quad \frac{\underline{} \div \underline{}}{\underline{} \div \underline{}} = \frac{\underline{}}{1} = \frac{\underline{}}{}$$

6. Austin's gas tank is $\frac{3}{4}$ full. How many times can he take a trip that requires $\frac{1}{8}$ of a tank of gas?

7. Nine tenths of the parking lot is left to shovel. Several people have each volunteered to shovel one tenth of the parking lot. How many volunteers are needed if they each do one tenth? (Notice that the denominators are the same. You do not need to use the Rule of Four.)

8. Mark has $\frac{5}{16}$ of a gallon of fuel left for his remote control airplane. The airplane's fuel tank holds $\frac{1}{16}$ of a gallon. How many times can Mark fill the fuel tank?

LESSON PRACTICE

Divide using the Rule of Four

1.

$$\frac{\underline{\quad}}{\underline{\quad}} \qquad \frac{2}{5} \div \frac{2}{3} \qquad \frac{\underline{\quad}}{\underline{\quad}} \qquad \frac{\underline{\quad} \div}{\underline{\quad} \div} \quad \underline{\quad} = \frac{\underline{\quad}}{1} = \quad \underline{\quad}$$

2.

$$\frac{\underline{\quad}}{\underline{\quad}} \qquad \frac{1}{2} \div \frac{1}{3} \qquad \frac{\underline{\quad}}{\underline{\quad}} \qquad \frac{\underline{\quad} \div}{\underline{\quad} \div} \quad \underline{\quad} = \frac{\underline{\quad}}{1} = \quad \underline{\quad}$$

3.

$$\frac{\underline{\quad}}{\underline{\quad}} \qquad \frac{2}{3} \div \frac{4}{5} \qquad \frac{\underline{\quad}}{\underline{\quad}} \qquad \frac{\underline{\quad} \div}{\underline{\quad} \div} \quad \underline{\quad} = \frac{\underline{\quad}}{1} = \quad \underline{\quad}$$

4.

$$\frac{\underline{\quad}}{\underline{\quad}} \qquad \frac{4}{8} \div \frac{1}{2} \qquad \frac{\underline{\quad}}{\underline{\quad}} \qquad \frac{\underline{\quad} \div}{\underline{\quad} \div} \quad \underline{\quad} = \frac{\underline{\quad}}{1} = \quad \underline{\quad}$$

5.

$$\frac{\underline{\quad}}{\underline{\quad}} \qquad \frac{1}{3} \div \frac{2}{5} \qquad \frac{\overline{\quad}}{\underline{\quad}} \qquad \frac{\underline{\quad}}{\underline{\quad}} \div \frac{\underline{\quad}}{\underline{\quad}} = \frac{\underline{\quad}}{1} = \frac{\underline{\quad}}{}$$

6. One third of the job remains to be done. How many people are needed to finish it if each is willing to do one ninth of the total job?

7. Kaylee has $\frac{4}{5}$ of a cake left. If she gives each of her friends $\frac{1}{10}$ of a cake, how many people can she serve?

8. Each board is $\frac{1}{6}$ of a foot thick. How many boards are there in a pile $\frac{1}{3}$ of a foot high?

SYSTEMATIC REVIEW

Divide. If the denominators are different, use the Rule of Four to make them the same.

1. $\dfrac{5}{6} \div \dfrac{1}{2} =$ _____

2. $\dfrac{4}{5} \div \dfrac{2}{5} =$ _____

3. $\dfrac{4}{5} \div \dfrac{1}{3} =$ _____

Multiply.

4. $\dfrac{3}{4} \times \dfrac{1}{8} =$ _____

5. $\dfrac{2}{3} \times \dfrac{1}{6} =$ _____

6. $\dfrac{6}{8} \times \dfrac{1}{4} =$ _____

Add or subtract.

7. $\dfrac{2}{3} - \dfrac{1}{6} =$ _____

8. $\dfrac{1}{2} - \dfrac{2}{5} =$ _____

9. $\dfrac{1}{2} + \dfrac{2}{3} + \dfrac{4}{5} =$ _____

Multiply. Estimate first if you wish.

10. $\begin{array}{r} 38 \\ \times\ 94 \\ \hline \end{array}$

11. $\begin{array}{r} 237 \\ \times\ 15 \\ \hline \end{array}$

12. $\begin{array}{r} 709 \\ \times\ 51 \\ \hline \end{array}$

QUICK REVIEW

Rounding and estimation are useful when performing long division. The purpose is to help you find the correct place value for your answer. For now, you do not have to be concerned about exact answers or remainders.

Round and estimate. The first one has been done for you.

13. $19\overline{)317} \rightarrow 20\overline{)300}$ with 10 above and $\underline{200}$ below

14. $51\overline{)498} \rightarrow$

15. $28\overline{)560} \rightarrow$

16. Phil and Lisa had $\frac{5}{8}$ of a pizza to share for lunch. What part of a pizza did each person get if they each had $\frac{1}{2}$ of what was there? (Each one gets $\frac{1}{2}$ **of** $\frac{5}{8}$, so we multiply.)

17. A piece of ribbon is $\frac{4}{6}$ of a yard long. How many $\frac{1}{3}$-yard pieces of ribbon can be cut? (The ribbon is being **divided** into pieces, so find $\frac{4}{6} \div \frac{1}{3}$.)

18. One half of the choir sang soprano, and one sixth sang alto. What part of the choir sang either soprano or alto? (We are **combining** sopranos and altos, so: $\frac{1}{2} + \frac{1}{6}$.)

SYSTEMATIC REVIEW

Divide. If the denominators are different, use the Rule of Four to make them the same.

1. $\dfrac{2}{3} \div \dfrac{1}{3} =$ _____

2. $\dfrac{3}{4} \div \dfrac{2}{5} =$ _____

3. $\dfrac{4}{6} \div \dfrac{1}{2} =$ _____

Multiply.

4. $\dfrac{4}{5} \times \dfrac{2}{5} =$ _____

5. $\dfrac{6}{8} \times \dfrac{1}{2} =$ _____

6. $\dfrac{1}{3} \times \dfrac{2}{5} =$ _____

Add or subtract.

7. $\dfrac{3}{4} - \dfrac{2}{3} =$ _____

8. $\dfrac{1}{10} + \dfrac{7}{8} =$ _____

9. $\dfrac{1}{2} + \dfrac{2}{9} + \dfrac{3}{4} =$ _____

Multiply. Estimate first if you wish.

10. $\begin{array}{r} 73 \\ \times\,28 \\ \hline \end{array}$

11. $\begin{array}{r} 829 \\ \times\,72 \\ \hline \end{array}$

12. $\begin{array}{r} 164 \\ \times\,53 \\ \hline \end{array}$

Round and estimate only. The first one has been done for you.

13. $5\,3\,\overline{)4\,2\,4} \rightarrow 5\,0\,\overline{)\begin{array}{c}8\\4\,0\,0\\\underline{4\,0\,0}\end{array}}$

14. $7\,4\,\overline{)7\,1\,1} \rightarrow$

15. $2\,2\,\overline{)8\,9\,0} \rightarrow$

Study the language used for each kind of fraction word problem.

16. Rod earned $50 and put one half of it in the bank. How many dollars did he put in the bank? ($\frac{1}{2}$ of 50)

17. One seventh of the trees in Kent's orchard have golden delicious apples, and one sixth of the trees have red delicious apples. What part of his trees have either red or golden delicious apples? ($\frac{1}{7} + \frac{1}{6}$)

18. Gina weeded $\frac{2}{5}$ of the garden, and Mark weeded $\frac{1}{3}$ of it. How much more did Gina weed than Mark weeded? ($\frac{2}{5} - \frac{1}{3}$)

19. The boss saw that $\frac{3}{4}$ of the job remained to be done. If each person does $\frac{1}{8}$ of the original job, how many people will be needed to finish the job? ($\frac{3}{4} \div \frac{1}{8}$)

20. Isabella is making a recipe that calls for $\frac{3}{4}$ of a cup of milk. She wants to make only $\frac{1}{2}$ of the recipe. How much milk should she use? ($\frac{1}{2} \times \frac{3}{4}$)

Divide. If the denominators are different, use the Rule of Four first to make them the same.

1. $\dfrac{5}{8} \div \dfrac{6}{8} =$ ──

2. $\dfrac{2}{3} \div \dfrac{1}{4} =$ ──

3. $\dfrac{2}{4} \div \dfrac{1}{6} =$ ──

Multiply.

4. $\dfrac{4}{5} \times \dfrac{1}{10} =$ ──

5. $\dfrac{5}{6} \times \dfrac{1}{12} =$ ──

6. $\dfrac{1}{2} \times \dfrac{1}{4} =$ ──

Add or subtract.

7. $\dfrac{1}{4} + \dfrac{2}{5} =$ ──

8. $\dfrac{4}{5} - \dfrac{7}{10} =$ ──

9. $\dfrac{6}{11} + \dfrac{1}{2} + \dfrac{2}{3} =$ ──

Multiply. Estimate first if you wish.

10. $\begin{array}{r} 35 \\ \times\ 16 \\ \hline \end{array}$

11. $\begin{array}{r} 182 \\ \times\ 68 \\ \hline \end{array}$

12. $\begin{array}{r} 390 \\ \times\ 41 \\ \hline \end{array}$

Round the divisors and dividends. Then find the quotient of the rounded values.

13. $4\,3\,\overline{)5\,5\,9}$ →

14. $7\,9\,\overline{)4\,1\,4}$ →

15. $1\,4\,\overline{)3\,9\,2}$ →

16. Nathan did $\frac{1}{5}$ of the job, but only $\frac{2}{3}$ of what he did was correct. What part of the job did he do correctly?

17. One fourth of a pie is left. If the leftover pie is divided into pieces that are each one sixteenth of a pie, how many people can be served?

18. Elspeth uses $\frac{3}{7}$ of her study time for math and $\frac{1}{4}$ of her study time for science. What part of her total study time is used for math and science?

19. Two hundred fifty-six swallows flew overhead. If each swallow ate thirty-five insects, how many insects were devoured?

20. One fifth of the pages in Warren's book had only pictures, and the rest had only words. What part of the pages had only words?

Warren has read five sixths of the pages with words. What part of the total number of pages has he read?

You learned in lesson 9 that a "fraction of a fraction" is a multiplication problem. You also learned that a "fraction of a number" means multiplication. Does that mean you should multiply whenever you see the word "of" in a problem?

Actually, almost every fraction word problem has the word "of" in it. How do you know when to multiply rather than add, subtract, or divide?

Every fraction is a fraction *of* something!

of an inch
of the people
of a pound
of the pie
of the job
of the garden

These phrases tell you what the problem is about, but they don't tell you what operation to use.

Fraction multiplication can be a fraction of a whole number or of another fraction.

$\frac{1}{4}$ of $\frac{3}{4}$
$\frac{1}{2}$ of 12
$\frac{2}{3}$ of $\frac{1}{2}$

These are solved by multiplication, but sometimes the value you are taking the "fraction of" is hidden in the text.

Study these multiplication problems from your student book to see how the value you are taking a "fraction of" may be in a different part of the sentence.

1. Joan picked $\frac{2}{3}$ of a bushel of tomatoes and gave $\frac{3}{5}$ of what she picked to her neighbor. What part of a whole bushel did her neighbor receive?

   "What she picked" is $\frac{2}{3}$ of a bushel, so the second fraction is a "fraction of" the first fraction. We multiply: $\frac{3}{5} \times \frac{2}{3} = \frac{6}{15}$. Notice that the answer is the same no matter which fraction you write first in the problem.

2. Ivan spotted 36 birds one morning. Three fourths of them were hawks. How many hawks did he see?

   "Them" refers to the birds, so the fraction is a "fraction of" the number of birds. Multiply: $\frac{3}{4} \times 36 = 27$

Underline the important words in each problem and connect the "fraction of" to the amount that you are finding the "fraction of." Write the multiplication problem.

3. Seven eighths of a pie was left over. Robin ate one seventh of what was left over. What part of a whole pie did she eat?

4. There were 30 days last month. Duncan woke up early on $\frac{1}{2}$ of the days last month. On how many days did Duncan wake up early?

5. Paul did $\frac{1}{5}$ of the job, but only $\frac{2}{3}$ of what he did was done correctly. What part of the job was done correctly?

This problem is a little different. Instead of finding one half of six days, we are multiplying one half of a mile by six to find the total distance.

6. Alan rode his bicycle one half of a mile every day for six days. How far did he ride his bicycle altogether?

Note: The Application and Enrichment pages for 11G include more tips for deciding what operation to use for fraction word problems. If you wish, you may complete those pages before beginning lesson 11.

Use the blocks to build rectangles. Find all the possible pairs of factors for each number and then list them in order from least to greatest. The first one has been done for you.

1. 16: <u>1</u> × <u>16</u> , <u>2</u> × <u>8</u> , <u>4</u> × <u>4</u>

   <u>1, 2, 4, 8,16</u>

2. 10: ___ × ___ , ___ × ___

   _____

3. 18: ___ × ___ , ___ × ___ , ___ × ___

   _____

Use the divisibility rules to answer the questions. Write *yes* or *no* in the blanks. Are these numbers divisible by 2?

4. 36 _____

5. 17_____

6. 125 _____

Are these numbers divisible by 10?

7. 41 _____

8. 100 _____

9. 30 _____

Are these numbers divisible by 5?

10. 15 _____

11. 30 _____

12. 351 _____

Are these numbers divisible by 3?

13. 48 _____

14. 65 _____

15. 105_____

Are these numbers divisible by 9?

16. 81_____

17. 234 _____

18. 67 _____

List all the factors for each number and then underline the common factors. Write the greatest common factor (GCF) for each pair of numbers in the blank. The factors for all of these numbers can be found with a set of blocks. The first one has been done for you.

19. 8: 1, 2, 4, 8

12: 1, 2, 3, 4, 6, 12

The GCF of 8 and 12 is 4 .

20. 15:

10:

The GCF of 15 and 10 is _____.

21. 6:

18:

The GCF of 6 and 18 is _____.

Use the blocks to build rectangles. Find all the possible pairs of factors for each number and then list them in order from least to greatest.

1. 9: ____ × ____ , ____ × ____

   _____

2. 4: ____ × ____ , ____ × ____

   _____

3. 14: ____ × ____ , ____ × ____

   _____

Use the divisibility rules to answer the questions. Write *yes* or *no* in the blanks.

4. Is 28 divisible by 2? _____

5. Is 460 divisible by 10? _____

6. Is 18 divisible by 5? _____

7. Is 138 divisible by 3? _____

8. Is 71 divisible by 3? _____

9. Is 853 divisible by 9? _____

For #10–11: Use the divisibility rules to find the pairs of factors for each number. Then list the factors in order from least to greatest. The first one has been done for you. You do not need to write out the steps to the right of the factors.

10. 36: _1_ × _36_   _1, 2, 3, 4, 6, 9, 12, 18, 36_

        _2_ × _18_   2 is a factor, and $36 \div 2 = 18$.

        _3_ × _12_   3 is a factor, and $36 \div 3 = 12$.

        _4_ × _9_   4 is a factor, and $36 \div 4 = 9$.

        _6_ × _6_   6 is a factor, and $36 \div 6 = 6$.

Nine is a factor, but it has already been used in 4 × 9. Therefore, the list of factors is complete.

11. 28: ___ × ___ , ___ × ___ , ___ × ___

_____

List all the factors for each number and then underline the common factors. Write the greatest common factor (GCF) for each pair of numbers in the blank.

12. 9:

    15:

    The GCF of 9 and 15 is ____.

13. 10:

    25:

    The GCF of 10 and 25 is ____.

14. 6:

    12:

    The GCF of 6 and 12 is ____.

Use the blocks to build rectangles. Find all the possible pairs of factors for each number and then list them in order from least to greatest.

1. 8: ___ × ___ , ___ × ___

    _____

2. 20: ___ × ___ , ___ × ___ , ___ × ___

    _____

3. 22: ___ × ___ , ___ × ___

    _____

Use the divisibility rules to answer the questions. Write *yes* or *no* in the blanks.

4. Is 47 divisible by 2? _____

5. Is 321 divisible by 10? _____

6. Is 35 divisible by 5? _____

7. Is 333 divisible by 3? _____

8. Is 103 divisible by 3?_____

9. Is 951 divisible by 9? _____

Use the divisibility rules to find the pairs of factors for each number. List the factors in order from least to greatest.

10. 42: ___ × ___ , ___ × ___ , ___ × ___ , ___ × ___

    _____

11. 34: ___ × ___ , ___ × ___

    _____

12. 50: ___ × ___ , ___ × ___ , ___ × ___

    _____

List all the factors for each number and then underline the common factors. Write the greatest common factor (GCF) for each pair of numbers in the blank.

13. 8:

    48:

    The GCF of 8 and 48 is _____.

14. 15:

    35:

    The GCF of 15 and 35 is _____.

15. 9:

    18:

    The GCF of 9 and 18 is _____.

Use the divisibility rules to answer the questions. Write yes or no in the blanks.

1. Is 362 divisible by 2?____

2. Is 370 divisible by 10? ____

3. Is 558 divisible by 5? ____

List all the factors for each number and then underline the common factors. Write the greatest common factor (GCF) for each pair of numbers in the blank.

4. 28:

   42:

   The GCF of 28 and 42 is ___.

5. 18:

   81:

   The GCF of 18 and 81 is ___.

Multiply.

6. $\dfrac{1}{2} \times \dfrac{1}{3}$

7. $\dfrac{2}{3} \times \dfrac{4}{5}$

8. $\dfrac{3}{6} \times \dfrac{2}{5}$

Divide using the Rule of Four.

9. $\dfrac{4}{5} \div \dfrac{1}{7}$

10. $\dfrac{5}{8} \div \dfrac{1}{3}$

11. $\dfrac{4}{5} \div \dfrac{1}{6}$

QUICK REVIEW

Here are the division problems you estimated in the last lesson. Compare the exact answers with the estimated answers from the last lesson if you wish.

Divide. Include a fraction in the answer if you are unable to divide evenly.

12.
$$16\frac{13}{19}$$
$$19\overline{)317}$$
$$\underline{19}$$
$$127$$
$$\underline{114}$$
$$13$$

13. $51\overline{)498}$

14. $28\overline{)560}$

15. Sammy ate $\frac{1}{2}$ of the cookies, and Jared finished off another $\frac{1}{4}$ of them. What part of the cookies was eaten? What part of the cookies is left?

16. One fifth of the swimmers in the class were beginners, and two sixths were at the intermediate level. What part of the class was either beginners or at the intermediate level?

17. Thirty-six students registered for an astronomy class. One third of the students were girls. How many girls and how many boys were registered for the class?

18. George worked harder than his friend and earned $\frac{4}{5}$ of the total pay for the job. The boss gave George $\frac{1}{2}$ of his share today. What part of the total pay does George get today?

Use the divisibility rules to answer the questions. Write *yes* or *no* in the blanks.

1. Is 264 divisible by 3? ____

2. Is 451 divisible by 9? ____

List all the factors for each number and then underline the common factors. Write the greatest common factor (GCF) for each pair of numbers in the blank.

3. 25:

   30:

   The GCF of 25 and 30 is ___.

4. 45:

   27:

   The GCF of 45 and 27 is ___.

Multiply.

5. $\dfrac{2}{3} \times \dfrac{7}{8}$

6. $\dfrac{1}{2} \times \dfrac{3}{9}$

7. $\dfrac{3}{4} \times \dfrac{5}{6}$

Divide using the Rule of Four.

8. $\dfrac{5}{8} \div \dfrac{1}{4}$

9. $\dfrac{6}{9} \div \dfrac{1}{6}$

10. $\dfrac{1}{2} \div \dfrac{3}{4}$

Use the Rule of Four to make denominators the same and then compare the fractions.

11. $\frac{3}{7}$ 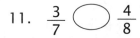 $\frac{4}{8}$ 12. $\frac{2}{5}$ 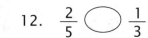 $\frac{1}{3}$

13. $\frac{5}{9}$ ◯ $\frac{3}{4}$

Divide. Include a fraction in the answer if you are unable to divide evenly.

14. $53\overline{)424}$ 15. $74\overline{)711}$

16. $22\overline{)890}$

17. Twenty-five buses drove into the stadium parking lot. Each bus carried 52 people. How many people came to the game on a bus?

18. Kelly did $\frac{1}{4}$ of the job on Monday, $\frac{1}{8}$ of it on Tuesday, and $\frac{1}{2}$ of it on Wednesday. What part of the job has she completed? What part of the job remains to be finished?

19. A race track that was $\frac{3}{4}$ of a mile long was divided by markers into sections that were $\frac{1}{8}$ of a mile long. Into how many sections was the track divided?

20. Kathy spent $\frac{1}{12}$ of this year traveling. One fourth of that time was spent on a ship. What part of this year did she spend traveling on a ship?

Use the divisibility rules to answer the questions. Write *yes* or *no* in the blanks.

1. Is 85 divisible by 2?____

2. Is 56 divisible by 5?____

List all the factors for each number and then underline the common factors. Write the greatest common factor (GCF) for each pair of numbers in the blank.

3. 16:

    34:

    The GCF of 16 and 34 is ___.

4. 12:

    40:

    The GCF of 12 and 40 is ___.

Multiply.

5. $\dfrac{5}{9} \times \dfrac{1}{2} =$

6. $\dfrac{3}{5} \times \dfrac{1}{4} =$

7. $\dfrac{6}{7} \times \dfrac{2}{3} =$

Divide using the Rule of Four.

8. $\dfrac{1}{9} \div \dfrac{1}{3} =$

9. $\dfrac{5}{6} \div \dfrac{2}{3} =$

10. $\dfrac{4}{8} \div \dfrac{1}{4} =$

Use the Rule of Four to make denominators the same and then compare the fractions.

11. $\frac{1}{2}$ ◯ $\frac{5}{6}$

12. $\frac{4}{8}$ ◯ $\frac{5}{10}$

13. $\frac{1}{12}$ ◯ $\frac{2}{9}$

Divide. Include a fraction in the answer if you are unable to divide evenly.

14. $43\overline{)559}$

15. $79\overline{)414}$

16. $14\overline{)392}$

17. A square room has 14 foot sides. The total length of all the door openings is $\frac{1}{7}$ of the perimeter. How many feet of baseboard are needed for the room? (Hint: First find the perimeter and then find how many feet represent door openings. Subtract that amount from the perimeter.)

18. In order for a number to be divisible by six, it must be divisible by both two and three. Is 114 divisible by six?

19. John budgeted money to take his children on an outing. He spent $\frac{1}{8}$ of the money on ice cream and $\frac{1}{4}$ of it on tickets to a museum. What part of his money did John spend on ice cream and tickets?

20. John budgeted $48 for the outing (#19). How much did he have left to spend after buying ice cream and museum tickets?

APPLICATION AND ENRICHMENT

You learned that *multiplying* a number by a fraction that is less than one gives an answer that is less than the original amount. On the other hand, *dividing* a number by a fraction less than one gives an answer that is greater than the original number.

Study the drawings and think about the meaning of each problem.

Example 1

Jim has 16 dollars. If he spends $\frac{1}{4}$ of his money on each gift, how much will he spend on each gift? $\frac{1}{4} \times \$16 = \$4$

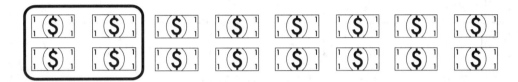

Example 2

Jim has 16 dollars. How many gifts can he buy that cost $\frac{1}{4}$ of a dollar apiece?
$\frac{16}{1} \div \frac{1}{4} = \frac{64}{4} \div \frac{1}{4} = 64$ gifts

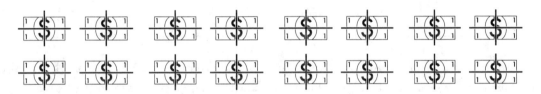

In Example 1, Jim spends $\frac{1}{4}$ of his money. The value of the money he started with is $16. Multiply to find a fraction of a number. If Jim spends one fourth of his money on each gift, he will spend less than the whole amount of his money on each one.

In Example 2, Jim started by dividing each of his dollars into fourths (quarters). Dividing 16 dollars by $\frac{1}{4}$ gives a total of 64 fourths or quarters. It makes sense that he would have more quarters than dollars.

Example 3

A rope is $\frac{4}{6}$ of a yard long. It is cut into pieces that are each $\frac{1}{3}$ of a yard long. How many pieces result? $\frac{4}{6} \div \frac{1}{3} = \frac{12}{18} \div \frac{6}{18} = 2$ pieces

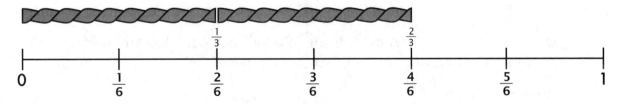

Although both fractions that you started with are less than one, the number of pieces is more than one.

When solving fraction word problems, you should look for *key ideas*, not just key words.

ADD	SUBTRACT
combining	difference
two parts together	how much more
total	how much less
altogether	what is left
MULTIPLY	DIVIDE
fraction of a fraction	how many pieces
fraction of a number	how many times
	something divided or cut into pieces
	how many people or parts are needed

Read carefully for meaning and build or draw a diagram to help you "see" what is happening in the problem. The problems below may not use the exact language shown in the boxes above, but they use similar ideas.

1. Pam has 3 pounds of raisins. How many times can she make a recipe that calls for $\frac{1}{6}$ of a pound of raisins?

2. Chris and Ryan had $\frac{3}{4}$ of a pizza to share for lunch. They each ate $\frac{1}{2}$ of what was there. What part of a whole pizza did each one eat?

3. One half of the cookies had pink icing, and one fourth of them had blue icing. The rest of the cookies were plain. How many cookies had icing?

4. One third of a pie is left over. If the leftover pie is divided among four people, what part of a pie will each person have?

5. One half of the fence is left to be painted. How many people are needed if each person will paint one fourth of the whole fence?

We encourage you to make up your own word problems to express these different ideas. Solve your problems and see if the answers make sense.

LESSON PRACTICE

Fill in the blanks in each equation to show what happens when the horizontal overlay is removed. The first one has been done for you.

1. ÷ 1 =

$$\frac{4 \div 4}{16 \div 4} = \frac{1}{4}$$

2. ÷ 1 =

$$\frac{\quad \div \quad}{\quad \div \quad} = \underline{\quad}$$

3. ÷ 1 =

$$\frac{\quad \div \quad}{\quad \div \quad} = \underline{\quad}$$

4. ÷ 1 =

$$\frac{\quad \div \quad}{\quad \div \quad} = \underline{\quad}$$

Simplify each fraction by dividing the numerator and denominator by the common factor. The first one has been done for you.

5. $\dfrac{6 \div 3}{15 \div 3} = \dfrac{2}{5}$.

6. $\dfrac{6 \div 2}{8 \div 2} = $ _____

7. $\dfrac{4 \div 4}{8 \div 4} = $ _____

Find the GCF (greatest common factor) of the numerator and denominator and use it to simplify each fraction. The first one has been done for you.

8. $\dfrac{8 \div 2}{10 \div 2} = \dfrac{4}{5}$

8: 1, 2, 4, 8

10: 1, 2, 5, 10

The GCF of 8 and 10 is 2 .

9. $\dfrac{4 \div }{24 \div } = $ _____

4:

24:

The GCF of 4 and 24 is ___ .

10. $\dfrac{6 \div }{18 \div } = $ _____

6:

18:

The GCF of 6 and 18 is ___ .

11. $\dfrac{18 \div }{30 \div } = $ _____

18:

30:

The GCF of 18 and 30 is ___ .

Fill in the blanks to show what happens when the horizontal overlay is removed.

1. ÷ =

$$\frac{\quad ÷ \quad}{\quad ÷ \quad} = \underline{\quad}$$

2. ÷ =

$$\frac{\quad ÷ \quad}{\quad ÷ \quad} = \underline{\quad}$$

3. ÷ =

$$\frac{\quad ÷ \quad}{\quad ÷ \quad} = \underline{\quad}$$

4. ÷ =

$$\frac{\quad ÷ \quad}{\quad ÷ \quad} = \underline{\quad}$$

Simplify each fraction by dividing the numerator and denominator by the common factor.

5. $\dfrac{5 \div 5}{20 \div 5} = $ _____

6. $\dfrac{10 \div 2}{12 \div 2} = $ _____

7. $\dfrac{12 \div 6}{18 \div 6} = $ _____

Find the GCF (greatest common factor) of the numerator and denominator and use it to simplify each fraction.

8. $\dfrac{3 \div}{15 \div} = $ _____

3:

15:

The GCF of 3 and 15 is ___ .

9. $\dfrac{8 \div}{12 \div} = $ _____

8:

12:

The GCF of 8 and 12 is ___ .

10. $\dfrac{15 \div}{18 \div} = $ _____

15:

18:

The GCF of 15 and 18 is ___ .

11. $\dfrac{14 \div}{21 \div} = $ _____

14:

21:

The GCF of 14 and 21 is ___ .

LESSON PRACTICE

Fill in the blanks to show what happens when the horizontal overlay is removed.

1.

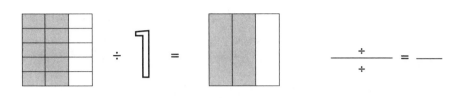

$$\frac{\quad \div \quad}{\quad \div \quad} = \underline{\quad}$$

2.

$$\frac{\quad \div \quad}{\quad \div \quad} = \underline{\quad}$$

Simplify each fraction by dividing the numerator and denominator by the common factor.

3. $\dfrac{27 \div 3}{33 \div 3} = \underline{\qquad}$ 4. $\dfrac{20 \div 4}{36 \div 4} = \underline{\qquad}$

5. $\dfrac{42 \div 6}{48 \div 6} = \underline{\qquad}$

Find the GCF (greatest common factor) of the numerator and denominator and use it to simplify each fraction.

6. $\dfrac{10 \div}{15 \div} =$ ——

7. $\dfrac{16 \div}{24 \div} =$ ——

10:

16:

15:

24:

The GCF of 10 and 15 is ___ .

The GCF of 16 and 24 is ___ .

Find the GCF of the numerator and denominator and use it to simplify each fraction. You may know the common factors for some of these without writing out all the factors for each number. If the number you divide by is not the *greatest* common factor, you will be able to divide again.

8. $\dfrac{25 \div}{30 \div} =$ ——

9. $\dfrac{6 \div}{8 \div} =$ ——

10. $\dfrac{10 \div}{16 \div} =$ ——

11. $\dfrac{18 \div}{21 \div} =$ ——

12. $\dfrac{45 \div}{50 \div} =$ ——

13. $\dfrac{27 \div}{36 \div} =$ ——

14. There are 56 people in the room, and $\frac{7}{14}$ of them have brown hair. Simplify the fraction. Then use the simplified fraction to find how many people have brown hair.

SYSTEMATIC REVIEW

12D

Find the GCF of the numerator and denominator and use it to simplify each fraction.

1. $\dfrac{12 \div}{28 \div} = \underline{\hspace{1cm}}$

2. $\dfrac{14 \div}{49 \div} = \underline{\hspace{1cm}}$

3. $\dfrac{35 \div}{50 \div} = \underline{\hspace{1cm}}$

Use the divisibility rules to answer the questions. Write *yes* or *no* in the blanks.

4. Is 195 divisible by 5? _____

5. Is 504 divisible by 9? _____

Follow the signs and then use the GCF to simplify your answers if possible.

6. $\dfrac{3}{4} + \dfrac{1}{8} = \underline{\hspace{1cm}} = \underline{\hspace{1cm}}$

7. $\dfrac{1}{7} + \dfrac{2}{5} = \underline{\hspace{1cm}} = \underline{\hspace{1cm}}$

8. $\dfrac{3}{8} \times \dfrac{2}{6} = \underline{\hspace{1cm}} = \underline{\hspace{1cm}}$

9. $\dfrac{4}{5} \times \dfrac{1}{10} = \underline{\hspace{1cm}} = \underline{\hspace{1cm}}$

10. $\dfrac{1}{4} \div \dfrac{5}{8} = \underline{\hspace{1cm}} = \underline{\hspace{1cm}}$

11. $\dfrac{2}{5} \div \dfrac{2}{3} = \underline{\hspace{1cm}} = \underline{\hspace{1cm}}$

QUICK REVIEW

When multiplying multiple-digit numbers, it is important to remember place value, to keep numbers lined up in the proper columns, and to regroup as necessary.

Multiply. Rewriting these problems on lined paper turned sideways may help you keep the columns aligned.

12.
$$\begin{array}{r} 5\ 3\ 1 \\ \times\ 6\ 2\ 4 \\ \hline \end{array}$$

13.
$$\begin{array}{r} 3,7\ 2\ 8 \\ \times\ 1\ 2\ 8 \\ \hline \end{array}$$

14.
$$\begin{array}{r} 6,5\ 9\ 3 \\ \times\ 7\ 5\ 6 \\ \hline \end{array}$$

15. Sally has $\frac{2}{4}$ of an orange. Simplify the fraction to find how many halves of an orange she has.

16. Four sixths of the work is left to do. Terry has assigned one sixth of the work to each of his helpers. If that will finish the work, how many helpers does Terry have?

17. Three fifths of the letters on the page were vowels. Of the vowels, two thirds were the letter *e*. What part of the letters on the page were *e*'s?

   (Simplify your answer).

18. Oscar finished $\frac{2}{3}$ of his math page before lunch. After lunch, he was able to finish another $\frac{1}{5}$ of the page. What part of Oscar's math page is now finished?

Find the GCF of the numerator and denominator and use it to simplify each fraction.

1. $\dfrac{24 \div}{30 \div} = \underline{\quad\quad}$

2. $\dfrac{18 \div}{28 \div} = \underline{\quad\quad}$

3. $\dfrac{15 \div}{35 \div} = \underline{\quad\quad}$

Use the divisibility rules to answer the questions. Write *yes* or *no* in the blanks.

4. Is 77 divisible by 3? _____.

5. Is 45 divisible by 10? _____.

Follow the signs and then use the GCF to simplify your answers if possible.

6. $\dfrac{1}{2} + \dfrac{3}{8} = \underline{\quad\quad} = \underline{\quad\quad}$

7. $\dfrac{2}{5} + \dfrac{3}{10} = \underline{\quad\quad} = \underline{\quad\quad}$

8. $\dfrac{2}{4} \times \dfrac{5}{6} = \underline{\quad\quad} = \underline{\quad\quad}$

9. $\dfrac{6}{8} \times \dfrac{3}{4} = \underline{\quad\quad} = \underline{\quad\quad}$

10. $\dfrac{1}{6} \div \dfrac{5}{9} = \underline{\quad\quad} = \underline{\quad\quad}$

11. $\dfrac{1}{2} \div \dfrac{3}{4} = \underline{\quad\quad} = \underline{\quad\quad}$

Multiply. Use lined paper if you wish.

12. $\begin{array}{r} 728 \\ \times\ 165 \\ \hline \end{array}$

13. $\begin{array}{r} 2,192 \\ \times\ 864 \\ \hline \end{array}$

14. $\begin{array}{r} 8,651 \\ \times\ 549 \\ \hline \end{array}$

15. Dad said that Eric must mow $\frac{5}{15}$ of the lawn. Simplify the fraction so that it is easier for Eric to picture.

16. An office supply company sold 512 boxes of paper clips that contained 150 clips each. How many paper clips were sold?

17. Two thirds of the beads on Debbie's necklace were white, and one fifth of them were pink. What part of the beads was either white or pink?

18. Esther got paid every 30 days. How many paychecks would Esther receive in 360 days?

19. Kym needed $\frac{3}{4}$ of a yard of red felt to make Christmas decorations. Her sister gave her $\frac{1}{2}$ of what she needed. What part of a yard did her sister give Kym?

20. Use your answer from #19 and the information given in that problem to find the amount of red felt Kym should buy.

SYSTEMATIC REVIEW

Find the GCF of the numerator and denominator and use it to simplify each fraction.

1. $\dfrac{18 \div}{63 \div}$ = _____

2. $\dfrac{26 \div}{32 \div}$ = _____

3. $\dfrac{42 \div}{48 \div}$ = _____

Use the divisibility rules to answer the questions. Write *yes* or *no* in the blanks.

4. Is 32 divisible by 2? _____

5. Is 168 divisible by 3? _____

Follow the signs and then use the GCF to simplify your answers if possible.

6. $\dfrac{4}{8} + \dfrac{1}{9}$ = _____ = _____

7. $\dfrac{4}{6} + \dfrac{1}{5}$ = _____ = _____

8. $\dfrac{6}{7} \times \dfrac{2}{6}$ = _____ = _____

9. $\dfrac{4}{5} \times \dfrac{1}{4}$ = _____ = _____

10. $\dfrac{1}{4} \div \dfrac{7}{8}$ = _____ = _____

11. $\dfrac{3}{6} \div \dfrac{4}{6}$ = _____ = _____

Multiply. Use lined paper if you wish.

12.
$$\begin{array}{r} 371 \\ \times\,244 \\ \hline \end{array}$$

13.
$$\begin{array}{r} 5,970 \\ \times\,186 \\ \hline \end{array}$$

14.
$$\begin{array}{r} 7,035 \\ \times\,369 \\ \hline \end{array}$$

15. Five hundred people went to a concert, and $\frac{18}{30}$ of them thought the music was too loud. Simplify the fraction and then use the result to find out how many people thought the music was too loud.

16. Four fifteenths of the choir sang soprano, and one fourth of them sang bass. Were there more basses or sopranos in the choir?

17. A bookstore bought 16 dozen pencils with holiday messages printed on them. They were resold in packs of three. How many packs of pencils did the store have to sell?

18. The packs of pencils (#17) were sold for $\frac{3}{4}$ of a dollar ($0.75). Use the fraction to find how much money the store received if all the packs of pencils were sold.

19. A rope is $\frac{4}{6}$ of a yard long. How many $\frac{1}{3}$-yard long pieces of rope may be cut from it?

20. One third of a pie is left over. It is to be divided among four people. What part of a pie will each person receive? Remember that four can be written as $\frac{4}{1}$. Divide to solve ($\frac{1}{3} \div \frac{4}{1}$).

You know how to compare fractions using either the overlays or the Rule of Four. You should also be using your knowledge and insight from working with different fractions to help you compare them. For example, if you eat $\frac{1}{4}$ of a pie in the morning and $\frac{1}{3}$ of the same pie in the evening, the amount of pie you ate is more than $\frac{1}{4}$ and also more than $\frac{1}{3}$. Is a guess of $\frac{1}{6}$ a reasonable answer?

$$\frac{1}{4} + \frac{1}{3} = \frac{1}{6} \qquad\qquad \textbf{false!}$$

Think: "If I cut the pie into six pieces, each piece will be smaller than $\frac{1}{4}$ or $\frac{1}{3}$. Adding $\frac{1}{4}$ and $\frac{1}{3}$ cannot possibly give an answer less than either of those fractions. Therefore, the answer cannot be $\frac{1}{6}$."

Use the overlays if necessary to see that $\frac{1}{6}$ is, in fact, less than either of the fractions you are adding. An answer to an addition problem involving fractions cannot be less than either one of the addends. Use the Rule of Four to compare: $\frac{3}{12} + \frac{4}{12} = \frac{7}{12}$ and $\frac{7}{12} > \frac{1}{6}$

Study the addition problems. Use your knowledge of fractions to decide if the answer could be true or if it is definitely false. Circle your response. Then add to find the exact sum.

1. $\dfrac{1}{2} + \dfrac{2}{5} = \dfrac{3}{7}$ **could be true**   **definitely false**

2. $\dfrac{1}{4} + \dfrac{1}{2} = \dfrac{2}{5}$ **could be true**   **definitely false**

When subtracting one fraction from another, the result cannot be greater than the fraction that you started with. Use your knowledge of fractions to decide if the answer could be true or if it is definitely false. Circle your response. Subtract to find the exact difference.

3. $\dfrac{2}{5} - \dfrac{1}{4} = \dfrac{9}{10}$ **could be true**   **definitely false**

4. $\dfrac{2}{3} - \dfrac{1}{4} = \dfrac{5}{12}$ **could be true**   **definitely false**

Use your knowledge of fractions to solve the word problems.

5. Jill ran $\frac{1}{3}$ of a mile this morning and $\frac{1}{5}$ of a mile this afternoon. How far did she run altogether?

6. Jed walked $\frac{3}{4}$ of a mile. He turned around and walked $\frac{1}{2}$ of a mile back towards his starting point before stopping to rest. How far was he from his starting point when he stopped to rest?

Solve each problem. Match the answer to the drawing that best represents it. Then create your own word problem to fit each one.

7. $\dfrac{1}{2} \times \dfrac{2}{3} =$

A.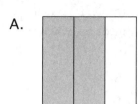

8. $\dfrac{1}{3} + \dfrac{1}{2} =$

B.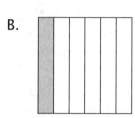

9. $\dfrac{1}{3} \div \dfrac{1}{2} =$

C.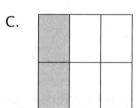

10. $\dfrac{1}{2} - \dfrac{1}{3} =$

D.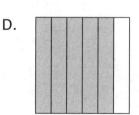

LESSON PRACTICE

Find the prime factors of each number using a factor tree. The first one has been done for you.

1. 63 = <u>3 × 3 × 7</u>

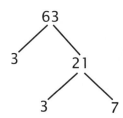

2. 45 = _____

3. 30 = _____

Find the prime factors of each number using repeated division. The first one has been done
for you.

4. 36 = <u>2 × 2 × 3 × 3</u>

```
      3
   3 | 9
  2 | 18
 2 | 36
```

5. 27 = _____

6. 50 = _____

Simplify the fractions using prime factors. The first one has been done for you.

7. $\dfrac{24}{32} = \dfrac{2\times2\times2\times3}{2\times2\times2\times2\times2} = \dfrac{2}{2} \times \dfrac{2}{2} \times \dfrac{2}{2} \times \dfrac{3}{2\times2} = 1 \times 1 \times 1 \times \dfrac{3}{2\times2} = \dfrac{3}{4}$

8. $\dfrac{20}{28} =$ —————— $=$ ——— $\times$ ——— $\times$ ——— $= 1 \times 1 \times$ ——— $=$ ———

9. $\dfrac{30}{50} =$ —————— $=$ ——— $\times$ ——— $\times$ ——— $= 1 \times 1 \times$ ——— $=$ ———

10. $\dfrac{21}{33} =$ —————— $=$ ——— $\times$ ——— $= 1 \times$ ——— $=$ ———

11. $\dfrac{18}{26} =$ —————— $=$ ——— $\times$ ——— $= 1 \times$ ——— $=$ ———

12. $\dfrac{24}{54} =$ —————— $=$ ——— $\times$ ——— $\times$ ——— $= 1 \times 1 \times$ ——— $=$ ———

LESSON PRACTICE

Find the prime factors of each number using a factor tree.

1. 100 = _____

2. 32 = _____

3. 72 = _____

Find the prime factors of each number using repeated division.

4. 48 = _____

5. 28 = _____

6. 42 = _____

Simplify the fractions using prime factors.

7. $\dfrac{45}{75}$ = ——————— = —— × —— × —— = 1 × 1 × —— = ——

8. $\dfrac{42}{60}$ = ——————— = —— × —— × —— = 1 × 1 × —— = ——

Simplify the fractions using prime factors. Instead of writing "1" each time you have the same factors in the numerator and the denominator, you may just draw lines through the matching factors. The first one has been done for you.

9. $\dfrac{12}{63} = \dfrac{2\times2\times\cancel{3}}{3\times\cancel{3}\times7} = \dfrac{2\times2}{3\times7} = \dfrac{4}{21}$

10. $\dfrac{33}{44}$ = ——————— = ———

11. $\dfrac{40}{90}$ = ——————— = ——— = ———

12. $\dfrac{27}{48}$ = ——————— = ——— = ———

Find the prime factors of each number using a factor tree.

1. 20 = _____

2. 81 = _____

3. 52 =_____

Find the prime factors of each number using repeated division.

4. 66 =_____

5. 44 = _____

6. 80 = _____

Simplify the fractions using prime factors. Draw lines through the matching factors in the numerator and the denominator.

7. $\dfrac{20}{24}$ = —————————— = ————— = ———

8. $\dfrac{30}{36}$ = —————————— = ————— = ———

9. $\dfrac{48}{54}$ = —————————— = ————— = ———

10. $\dfrac{9}{27}$ = —————————— = —————

11. $\dfrac{15}{25}$ = —————————— = —————

12. $\dfrac{12}{15}$ = —————————— = ————— = ———

SYSTEMATIC REVIEW

Find the prime factors using the method you prefer.

1. 26 = _____

2. 60 = _____

3. 24 = _____

Simplify the fractions using prime factors.

4. $\dfrac{18}{24}$ = _____ = _____ = _____

5. $\dfrac{15}{30}$ = _____ = _____

Follow the signs and then simplify the answer if possible using the GCF or prime factors.

6. $\dfrac{1}{4} + \dfrac{2}{8}$ = _____ = _____

7. $\dfrac{5}{8} - \dfrac{1}{2}$ = _____ = _____

8. $\dfrac{2}{7} + \dfrac{1}{3}$ = _____ = _____

9. $\dfrac{1}{6} \div \dfrac{1}{2}$ = _____ = _____

10. $\dfrac{5}{9} \times \dfrac{3}{7}$ = _____ = _____

11. $\dfrac{8}{11} \times \dfrac{3}{4}$ = _____ = _____

QUICK REVIEW

When you divide multiple-digit numbers, you may need to use trial and error to find the proper answer for each step. Estimation can also help in this process. Be careful to keep numbers aligned in the proper columns (place value).

Divide. Include a fraction in the answer if you cannot divide evenly. Use lined paper turned sideways if you wish. The first one has been done for you.

12.

$$
\begin{array}{r}
72 \frac{19}{71} \\
71 \overline{\smash{\big)}5,131} \\
-497 \\
\hline
161 \\
-142 \\
\hline
19
\end{array}
$$

13. $53\overline{\smash{\big)}9,442}$

14. $189\overline{\smash{\big)}4,925}$

15. Joel found that the names for one sixth of the months of the year begin with the letter A. How many names begin with A?

16. Faith needs $\frac{3}{4}$ of a yard of blue calico and $\frac{1}{8}$ of a yard of green calico for her quilt. How many yards does she need in all? Simplify the answer to lowest terms.

17. Adam drove 2,845 miles in his truck. The truck used a gallon of diesel fuel for every five miles. How many gallons of fuel did Adam's truck use in all?

18. Five hundred twenty-eight mosquitoes laid their eggs. Each mosquito had 73 offspring that survived. How many new mosquitoes are there?

Find the prime factors using the method you prefer.

1. $81 = $ _____

2. $90 = $ _____

3. $75 = $ _____

Simplify the fractions using prime factors.

4. $\dfrac{8}{22} = $ _____ $= $ _____ $= $ _____

5. $\dfrac{32}{48} = $ _____ $= $ _____

Follow the signs and then simplify the answer if possible using the GCF or prime factors.

6. $\dfrac{1}{2} + \dfrac{2}{6} = $ _____ $= $ _____

7. $\dfrac{6}{9} - \dfrac{1}{3} = $ _____ $= $ _____

8. $\dfrac{3}{5} + \dfrac{1}{10} = $ _____ $= $ _____

9. $\dfrac{4}{16} \div \dfrac{3}{4} = $ _____ $= $ _____

10. $\dfrac{5}{9} \times \dfrac{2}{6} = $ _____ $= $ _____

11. $\dfrac{2}{5} \times \dfrac{3}{4} = $ _____ $= $ _____

Use the Rule of Four to make denominators the same and then compare the fractions.

12. $\dfrac{5}{8} \bigcirc \dfrac{4}{6}$

13. $\dfrac{1}{3} \bigcirc \dfrac{2}{7}$

14. $\dfrac{5}{10} \bigcirc \dfrac{4}{8}$

Divide. Include a fraction in the answer if you are unable to divide evenly. You may use lined paper if you wish.

15. $38\overline{)1{,}1\,3\,0}$

16. $22\overline{)2{,}6\,8\,6}$

17. $235\overline{)5{,}0\,3\,2}$

18. Lance put 45 pennies in a jar every day for a year. How many pennies did he have at the end of a year? (365 days)

19. Lynne practiced her clarinet for $\frac{1}{3}$ of an hour on Monday and $\frac{5}{12}$ of an hour on Tuesday. For what part of an hour has she practiced? (Do not simplify the answer.)

20. Use your answer from #19 to find how many minutes Lynne practiced. Now simplify your answer from #19 and find the number of minutes again using the simplified fraction. Which fraction did you find easier to use?

Find the prime factors using the method you prefer.

1. 64 = _____

2. 16 = _____

3. 45 = _____

Simplify the fractions using prime factors.

4. $\dfrac{81}{90}$ = _____ = _____ = ____

5. $\dfrac{12}{18}$ = _____ = _____

Follow the signs and then simplify the answer if possible using the GCF or prime factors.

6. $\dfrac{3}{8} + \dfrac{1}{6}$ = ____ = ____

7. $\dfrac{9}{10} - \dfrac{2}{5}$ = ____ = ____

8. $\dfrac{3}{6} + \dfrac{1}{4}$ = ____ = ____

9. $\dfrac{1}{5} \div \dfrac{4}{10}$ = ____ = ____

10. $\dfrac{4}{7} \times \dfrac{2}{8}$ = ____ = ____

11. $\dfrac{5}{12} \times \dfrac{3}{5}$ = ____ = ____

Use the Rule of Four to make denominators the same. Then compare the fractions.

12. $\dfrac{6}{11}$ $\bigcirc$ $\dfrac{7}{10}$

13. $\dfrac{1}{4}$ $\bigcirc$ $\dfrac{4}{9}$

14. $\dfrac{5}{12}$ $\bigcirc$ $\dfrac{1}{3}$

Divide. Include a fraction in the answer if you are unable to divide evenly.

15. $96\overline{)3{,}6\,2\,1}$

16. $73\overline{)7{,}1\,9\,2}$

17. $120\overline{)6{,}8\,3\,1}$

18. Lara has 2,075 pennies. If she spends 25¢ a day, how many days will it take to spend all her pennies?

19. Greg did $\frac{1}{3}$ of the job in the morning, $\frac{1}{6}$ of it after lunch, and $\frac{1}{8}$ of it in the evening. What part of the job has been completed? Give your answer as a simplified fraction.

20. Use your answer from #19 to find what part of the job is left to do.

In lesson 11, you learned how to find the greatest common factor (GCF) of two numbers by listing the factors of each number. In this chapter, you learned how to find prime factors using a factor tree or repeated division. You can also use prime factors to find the GCF of two numbers. First, note which prime factors are common to both numbers. Then multiply the common factors to find the GCF. Study the example.

Example 1
Use prime factors to find the GCF of 18 and 27.

The prime factors of 18 are $2 \times 3 \times 3$.
The prime factors of 27 are $3 \times 3 \times 3$. $\boxed{3 \times 3 = 9}$

Note carefully what happened in the example. The number 2 is not common to both answers, so we ignored the 2. Two 3s are common to both answers, so we put two 3s in the box. Multiplying the two common prime factors gives a result of 9, and, sure enough, 9 is the greatest common factor of 18 and 27.

1. Use prime factors to find the GCF of 66 and 84.

    66:

    84:

    Write the factors that are *common* to, or included in, both lists in the box. Multiply the factors you wrote in the box. Check to be sure that your answer is a factor of both 66 and 84.

2. Use prime factors to find the GCF of 62 and 93.

3. Use prime factors to find the GCF of 40 and 90.

Factors may be used to rewrite an addition problem as a multiplication problem. The *Distributive Property of Multiplication over Addition* allows us to multiply each number inside the parentheses in the new expression by the common factor to get back to the original problem. If you need to review how to work with parentheses, you can go to Application and Enrichment 3G and 6G.

Example 1
Rewrite 30 + 42 using the greatest common factor (GCF).

The GCF of 30 and 42 is 6.
Using 6 as a factor, you can rewrite 30 + 42 as 6(5 + 7).

Example 2
Solve 6(5 + 7) by adding and then multiplying. Use the Distributive Property to check your answer.

6(5 + 7) = 6(12) = 72.
Check: (6 × 5) + (6 × 7) = 30 + 42 = 72

Rewrite each problem using the greatest common factor. Add the numbers inside the parentheses and multiply by the GCF to solve each problem. Use the Distributive Property to check your answer.

4. 15 + 10 =

5. 18 + 24 =

6. 32 + 56 =

7. 45 + 81 =

8. 6 + 39 =

Count spaces to find the denominator and numerator and then write the length of each line. The drawings on this page are slightly longer than an inch to make them easier to read. The first one has been done for you.

1. 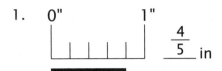 $\dfrac{4}{5}$ in

2. 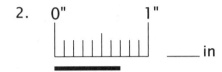 ___ in

3. 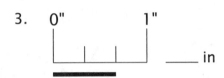 ___ in

4.  ___ in

Divide the line into equal parts. Then draw a line of the given length. The first one has been done for you.

5. 0"   1" $\dfrac{3}{4}$ in

6. 0"   1" $\dfrac{2}{5}$ in

Give the length of each line as sixteenths of an inch. Simplify if possible. If you cannot simplify, leave the second blank empty. The first one has been done for you.

7. 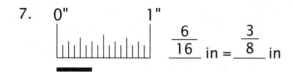 $\dfrac{6}{16}$ in = $\dfrac{3}{8}$ in

8. 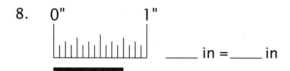 _____ in = _____ in

9.  _____ in = _____ in

10. 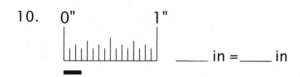 _____ in = _____ in

11. 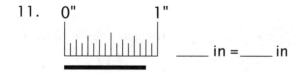 _____ in = _____ in

12. 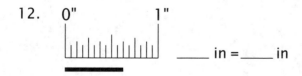 _____ in = _____ in

Count spaces to find the denominator and numerator and then write the length of each line. The drawings on these pages are slightly longer than an inch to make them easier to read.

1.  _____ in

2. 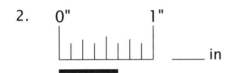 _____ in

3. 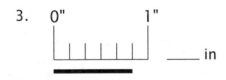 _____ in

4. 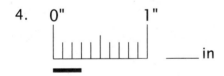 _____ in

Divide the line into equal parts. Draw a line of the given length.

5. 0"   1" $\frac{3}{6}$ in

6. 0"   1" $\frac{1}{4}$ in

Give the length of each line as sixteenths of an inch. Simplify if possible. If you cannot simplify, leave the second blank empty.

7.  _____ in = _____ in

8.  _____ in = _____ in

9. 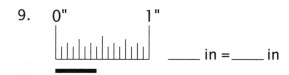 _____ in = _____ in

10. 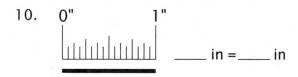 _____ in = _____ in

11. 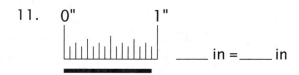 _____ in = _____ in

12. 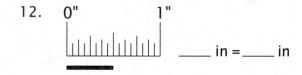 _____ in = _____ in

Count spaces to find the denominator and numerator and then write the length of each line. The drawings on these pages are slightly longer than an inch to make them easier to read.

1. 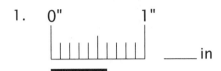 _____ in

2. 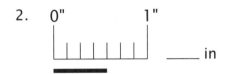 _____ in

3.  _____ in

4.  _____ in

Divide the line into equal parts. Draw a line of the given length.

5. 0"   1" $\frac{2}{7}$ in

6. 0"   1" $\frac{4}{5}$ in

Give the length of each line as sixteenths of an inch. Simplify if possible. If you cannot simplify, leave the second blank empty.

7. 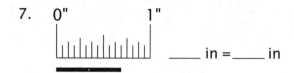 _____ in = _____ in

8.  _____ in = _____ in

9. 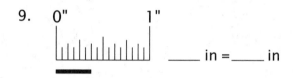 _____ in = _____ in

10. 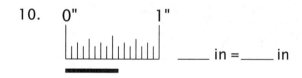 _____ in = _____ in

11. 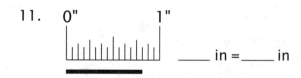 _____ in = _____ in

12. 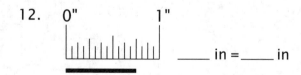 _____ in = _____ in

SYSTEMATIC REVIEW

Write the length of each line beside the ruler. Simplify if possible. If you cannot simplify, leave the second blank empty.

1. 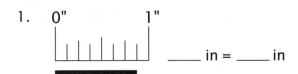 _____ in = _____ in

2.  _____ in = _____ in

3.  _____ in = _____ in

4. _____ in = _____ in

Find the prime factors using the method you prefer.

5. 28 = _____

6. 55 = _____

7. 84 = _____

Simplify the fractions using prime factors.

8. $\dfrac{48}{64}$ = _____ = _____ = _____

QUICK REVIEW

You have already reviewed perimeter. It is the distance around a figure and is found by adding the lengths of the sides. *Area* is the amount of space enclosed by the figure. Area is always given in square units. To find the area of a rectangle, multiply the length, or base, by the height.

Find the area of each rectangle. The first one has been done for you.

9.   3"

5"   (5)(3) = 15   A = __15 sq in__

This may also be written as 15 in^2. It is read as "15 square inches."

10.   10'

20'

A = _____

11.   7 yd

15 yd

A = _____

12.   Is 195 divisible by two?

13.   Is 183 divisible by three?

14.   What is the GCF of 12 and 18?

15.   What is the GCF of 25 and 50?

16.   The floor tile Patrick wants for his room is sold by the square foot. His room measures 15 feet by 13 feet. How many square feet of floor tile does he need? (area)

Write the length of each line beside the ruler. Simplify if possible. If you cannot simplify, leave the second blank empty.

1.  ____ in = ____ in

2. 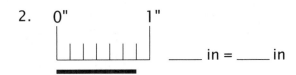 ____ in = ____ in

3. 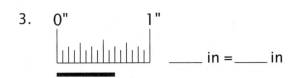 ____ in = ____ in

4. 0"   1"

   ____ in = ____ in

Find the prime factors using the method you prefer.

5. 70 = _____

6. 45 = _____

7. 30 = _____

Simplify the fractions using prime factors.

8. $\dfrac{33}{63}$ = _____ = _____ = _____

Find the area of each rectangle.

9. 5"   8"   A = _____

10. 45'   90'   A = _____

11. 15 yd   28 yd   A = _____

12. Is 263 divisible by nine?

13. What is the GCF of 21 and 42?

14. Sam used $\frac{1}{3}$ of his allowance for clothes. Of that amount, $\frac{3}{4}$ went for school clothes. What part of his allowance was used for school clothes? Simplify your answer.

15. Mom has a measuring cup that holds $\frac{2}{3}$ of a cup and one that holds $\frac{3}{4}$ of a cup. Compare fractions to see which holds more.

16. What is the difference between the measures of the two cups in #15?

17. Patrick wants to put a border with a sports theme all around the walls of his 15 ft by 13 ft room. How many feet of border does he need? (Don't worry about door and window openings.)

18. When Patrick put up his border (#17), he discovered he had only $\frac{3}{4}$ of what was needed. How many more feet does he need to buy?

Write the length of each line beside the ruler. Simplify if possible. If you cannot simplify, leave the second blank empty.

1.  ____ in = ____ in

2. 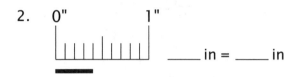 ____ in = ____ in

3. 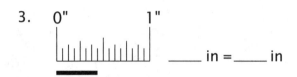 ____ in =____ in

4. ____ in =____ in

Find the prime factors using the method you prefer.

5. 66 = _____

6. 28 = _____

7. 54 = _____

Simplify the fractions using prime factors.

8. $\dfrac{75}{100}$ = ——————— = ——————— = ———

Find the area of each rectangle.

9. [rectangle] 2"   A = _____
    3"

10. [rectangle] 6'   A = _____
    11'

11. [rectangle] 19 yd   A = _____
    35 yd

12. Is 255 divisible by five?

13. What is the GCF of 24 and 36?

14. Sally and her friends ate three sevenths of the chocolates. Bill and his friends ate one sixth of them. What part of the chocolates has been eaten? Can this fraction be simplified?

15. A turtle set out to crawl $\frac{8}{12}$ of a mile. It stopped to rest every $\frac{1}{6}$ of a mile. How many times did the turtle stop to rest? (Your answer will include the final stop at the end of the trip.)

16. Farmer Brown packed his eggs into crates that held 36 dozen each. How many eggs were in each crate?

17. If Farmer Brown (#16) had 2,160 eggs, how many crates could he fill?

18. Meg plans to buy carpet for her room. Does she need to find the area or the perimeter of the floor?

APPLICATION AND ENRICHMENT

If you keep one factor of a multiplication problem the same and double the other factor, the product will double as well. Example 1 uses an area problem to illustrate this concept.

Example 1

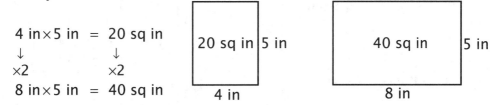

4 in × 5 in = 20 sq in
↓   ↓
×2   ×2
8 in × 5 in = 40 sq in

If you take 2 × 4, the new factor will be 8, and the new product that results will be 2 × 40 = 80.

Lisa represented a multiplication problem with factors of 2 and 3 by drawing a rectangle with one side 2 inches long and the other side 3 inches long. Answer the questions. Sketch the rectangles if you wish.

1. What is the product of the factors?

   2 in × 3 in = ____ sq in

2. Lisa took three times the first factor. Fill in the blanks.

   ____ in × 3 in = ____ sq in

3. The product of problem 2 is _____ times greater than the product of problem 1.

4. Lisa drew a rectangle to represent a new multiplication problem. The first factor is five times the first factor in problem 1. The second factor is the same as the second factor in problem 1. The new product is _____ times greater than the product of problem 1.

5. The first factor is ten times the first factor in problem 1. The second factor is the same as the second factor in problem 1. The new product is __ times greater than the product in problem 1.

This principle of re-sizing a rectangle by changing one factor also works when you multiply one factor by a fraction rather than a whole number.

Example 2

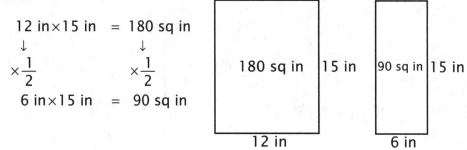

$$12 \text{ in} \times 15 \text{ in} = 180 \text{ sq in}$$
$$\downarrow \qquad\qquad \downarrow$$
$$\times\frac{1}{2} \qquad\qquad \times\frac{1}{2}$$
$$6 \text{ in} \times 15 \text{ in} = 90 \text{ sq in}$$

If you take $\frac{1}{2}$ of 6, the new factor will be 3, and the new product that results will be $\frac{1}{2}$ of 90 = 45.

Linda represented a multiplication problem with factors of 24 and 10 by drawing a rectangle with one side 24 inches long and the other side 10 inches long. Answer the questions. Sketch the rectangles if you wish.

6. What is the product of the factors?

    24 in × 10 in = _____ sq in

7. Linda found one half of the first factor. Fill in the blanks.

    _____ in × 10 in = _____ sq in

8. The product of problem 7 is _____ times as great as the product of problem 6.

9. Linda drew a rectangle to represent a new multiplication problem. The first factor is one sixth of the first factor in problem 6. The second factor is the same as the second factor in problem 6. The new product is _____ times as great as the product of problem 6.

10. The first factor is one fourth of the first factor in problem 6. The second factor is the same as the second factor in problem 6. The new product is _____ times as great as the product of problem 6.

Change each mixed number to an improper fraction. The first one has been done for you.

1.

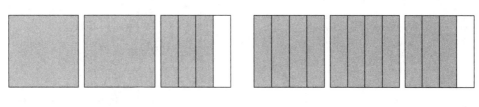

$$1 \quad + \quad 1 \quad + \quad \frac{3}{4} \quad = \quad \frac{4}{4} \quad + \quad \frac{4}{4} \quad + \quad \frac{3}{4} \quad = \quad \frac{11}{4}$$

2.

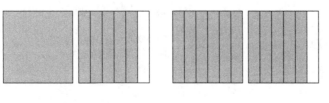

$$1 \quad + \quad \frac{5}{6} \quad = \quad \underline{\quad} \quad + \quad \underline{\quad} \quad = \quad \underline{\quad}$$

3.

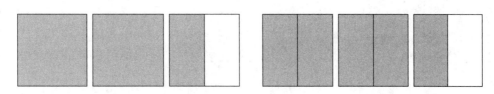

$$1 \quad + \quad 1 \quad + \quad \frac{1}{2} \quad = \quad \underline{\quad} \quad + \quad \underline{\quad} \quad + \quad \underline{\quad} \quad = \quad \underline{\quad}$$

Change each improper fraction to a mixed number. The first one has been done for you.

4.

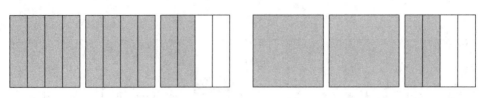

$$\frac{4}{4} \quad + \quad \frac{4}{4} \quad + \quad \frac{2}{4} \quad = \quad 1 \quad + \quad 1 \quad + \quad \frac{2}{4} \quad = 2\,\frac{2}{4}$$

5.

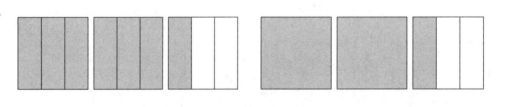

$$\frac{3}{3} \quad + \quad \frac{3}{3} \quad + \quad \frac{1}{3} \quad = \quad \underline{\quad\quad} \quad + \quad \underline{\quad\quad} \quad + \quad \underline{\quad\quad} \quad = \quad \underline{\quad\quad\quad}$$

6.

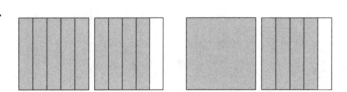

$$\frac{5}{5} \quad + \quad \frac{4}{5} \quad = \quad \underline{\quad\quad} \quad + \quad \underline{\quad\quad} \quad = \quad \underline{\quad\quad}$$

Change each mixed number to an improper fraction.

1.

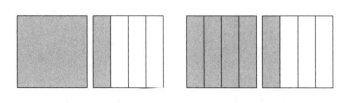

$$1 \quad + \quad \frac{1}{4} \quad = \quad \underline{\quad} \quad + \quad \underline{\quad} \quad = \quad \underline{\quad}$$

2.

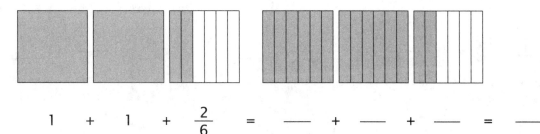

$$1 \quad + \quad 1 \quad + \quad \frac{2}{6} \quad = \quad \underline{\quad} \quad + \quad \underline{\quad} \quad + \quad \underline{\quad} \quad = \quad \underline{\quad}$$

Change each mixed number to an improper fraction. The first one has been done for you.

3. $2\frac{2}{4} = \frac{4}{4} + \frac{4}{4} + \frac{2}{4} = \frac{10}{4}$

4. $3\frac{2}{3} = \frac{}{3} + \frac{}{3} + \frac{}{3} + \frac{}{3} = \underline{\quad}$

Change each improper fraction to a mixed number.

5.

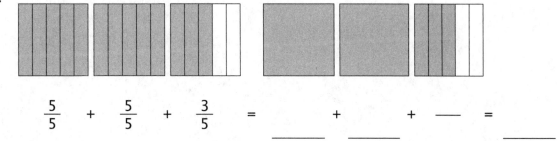

$$\frac{5}{5} \quad + \quad \frac{5}{5} \quad + \quad \frac{3}{5} \quad = \quad \underline{\quad} \quad + \quad \underline{\quad} \quad + \quad \underline{\quad} \quad = \quad \underline{\quad}$$

Convert each improper fraction to a mixed number. The first one has been done for you.

6. $\dfrac{14}{5} = \dfrac{5}{5} + \dfrac{5}{5} + \dfrac{4}{5} = 1 + 1 + \dfrac{4}{5} = 2\dfrac{4}{5}$ _____

7. $\dfrac{3}{2} = \dfrac{}{2} + \dfrac{}{2} = \text{\_\_\_\_} + \dfrac{}{2} = \text{\_\_\_\_}$

8. $\dfrac{23}{6} = \dfrac{}{6} + \dfrac{}{6} + \dfrac{}{6} + \dfrac{}{6} = \text{\_\_\_\_} + \text{\_\_\_\_} + \text{\_\_\_\_} + \dfrac{}{6} = \text{\_\_\_\_}$

LESSON PRACTICE

Convert each mixed number to an improper fraction.

1.

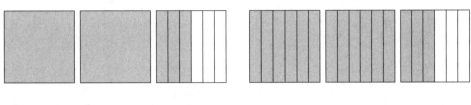

$$1 \quad + \quad 1 \quad + \quad \frac{3}{6} \quad = \quad \text{—} \quad + \quad \text{—} \quad + \quad \text{—} \quad = \quad \text{—}$$

2. $2\dfrac{1}{2} = \dfrac{}{2} + \dfrac{}{2} + \dfrac{}{2} = \text{———}$

3. $1\dfrac{4}{5} = \dfrac{}{5} + \dfrac{}{5} = \text{———}$

4. $3\dfrac{5}{8} = \dfrac{}{8} + \dfrac{}{8} + \dfrac{}{8} + \dfrac{}{8} = \text{———}$

5. $2\dfrac{1}{4} = \dfrac{}{4} + \dfrac{}{4} + \dfrac{}{4} = \text{———}$

Change the improper fraction to a mixed number.

6.

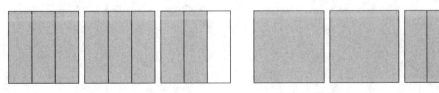

$$\frac{3}{3} \quad + \quad \frac{3}{3} \quad + \quad \frac{2}{3} \quad = \quad \underline{\quad} \quad + \quad \underline{\quad} \quad + \quad \text{—} \quad = \quad \underline{\quad\quad}$$

Convert each improper fraction to a mixed number. You do not need to write "1" each time. The first one has been done for you.

7. $\dfrac{11}{4} = \dfrac{4}{4} + \dfrac{4}{4} + \dfrac{3}{4} = 2\dfrac{3}{4}$

8. $\dfrac{8}{7} = \dfrac{}{7} + \dfrac{}{7} =$ _____

9. $\dfrac{5}{2} = \dfrac{}{2} + \dfrac{}{2} + \dfrac{}{2} =$ _____

10. $\dfrac{17}{4} = \dfrac{}{4} + \dfrac{}{4} + \dfrac{}{4} + \dfrac{}{4} + \dfrac{}{4} =$ _____

Convert each mixed number to an improper fraction.

1. $2\dfrac{2}{3} = \dfrac{}{3} + \dfrac{}{3} + \dfrac{}{3} = $ _____

2. $3\dfrac{1}{2} = \dfrac{}{2} + \dfrac{}{2} + \dfrac{}{2} + \dfrac{}{2} = $ _____

Convert each improper fraction to a mixed number.

3. $\dfrac{13}{9} = \dfrac{}{9} + \dfrac{}{9} = $ _____

4. $\dfrac{11}{5} = \dfrac{}{5} + \dfrac{}{5} + \dfrac{}{5} = $ _____

Write the length of each line beside the ruler. Simplify if possible. If the fraction cannot be simplified, leave the second blank empty.

5. 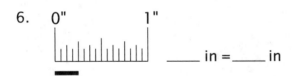 _____ in = _____ in

6. 0"   1" _____ in = _____ in

Simplify the fractions using any method you wish. If you divide by a factor that is not the GCF, you will be able to divide again. You can simplify in two steps, but be sure the final result is simplified completely.

7. $\dfrac{16}{24} = $ _____

8. $\dfrac{28}{35} = $ _____

9. $\dfrac{54}{63} = $ _____

QUICK REVIEW

Multiply length by width to find the area of a square. The answer is always given in square units. Remember that you add the lengths of all the sides to find the perimeter.

Find the area and perimeter of each square. The first two have been done for you.

8'
8'

10. A = ___64 sq ft___

(8 ft)(8 ft) = 64 sq ft

11. P = ___32 ft___

8 ft + 8 ft + 8 ft + 8 ft = 32 ft

10"
10"

12. A = _____

13. P = _____

5 miles
5 miles

14. A = _____

15. P = _____

16. Find the prime factors of 93.

17. What is the GCF of 44 and 55?

18. A path that is $\frac{7}{8}$ of a mile long is divided into sections that are $\frac{2}{16}$ of a mile long. How many sections are created? Simplify $\frac{2}{16}$ first to make your work easier.

Convert each mixed number to an improper fraction.

1. $1\dfrac{5}{6} = \dfrac{}{6} + \dfrac{}{6} = \underline{}$

2. $3\dfrac{4}{9} = \dfrac{}{9} + \dfrac{}{9} + \dfrac{}{9} + \dfrac{}{9} = \underline{}$

Convert each improper fraction to a mixed number.

3. $\dfrac{15}{8} = \dfrac{}{8} + \dfrac{}{8} = \underline{}$

4. $\dfrac{33}{10} = \dfrac{}{10} + \dfrac{}{10} + \dfrac{}{10} + \dfrac{}{10} = \underline{}$

Write the length of each line beside the ruler. Simplify if possible. If the fraction cannot be simplified, leave the second blank empty.

5. 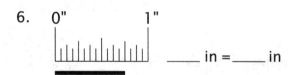

   0"   1"

   _____ in = _____ in

6. 0"   1"

   _____ in = _____ in

Simplify the fractions using any method you wish.

7. $\dfrac{27}{45} = \underline{}$ 8. $\dfrac{44}{60} = \underline{}$

9. $\dfrac{24}{30} = \underline{}$

QUICK TIP

The expression 8^2 means 8×8 and is read "eight squared." If you know the length of one side of a square, you can square the number to find the area of the square.

Find the area of a square with the given side. The first one has been done for you.

10. S = 7 feet   A = <u>49 sq ft or ft^2</u>

$7^2 = 7$ ft $\times 7$ ft $= 49$ ft^2

11. S = 11 miles   A = _____

12. S = 14 inches   A = _____

13. Find the prime factors of 81.

14. What is the GCF of 18 and 28?

15. Is 95 divisible by two?

16. One third of the people in the room have blue eyes. Five sixths of the blue-eyed people also have blond hair. What part of the people in the room are blue-eyed blonds?

17. Tim earns $35 a day. How much will he earn in 175 days?

18. What is the perimeter of a square that measures 12 miles on each side?

Convert each mixed number to an improper fraction.

1. $1\frac{3}{8} = \frac{}{8} + \frac{}{8} = \underline{\qquad}$

2. $3\frac{2}{7} = \frac{}{7} + \frac{}{7} + \frac{}{7} + \frac{}{7} = \underline{\qquad}$

Convert each improper fraction to a mixed number.

3. $\frac{11}{7} = \frac{}{7} + \frac{}{7} = \underline{\qquad}$

4. $\frac{28}{6} = \frac{}{6} + \frac{}{6} + \frac{}{6} + \frac{}{6} + \frac{}{6} = \underline{\qquad}$

Write the length of each line beside the ruler. Simplify if possible. If the fraction cannot be simplified, leave the second blank empty.

5. 0"   1"

_____ in = _____ in

6. 0"   1"

_____ in = _____ in

Simplify the fractions using any method you wish.

7. $\frac{50}{70} = \underline{\qquad}$ 8. $\frac{24}{36} = \underline{\qquad}$

9. $\frac{3}{9} = \underline{\qquad}$

Find the area of a square with the given side.

10. S = 25 feet   A = _____

11. S = 2 inches   A = _____

12. S = 17'   A = _____

Solve. The first one has been done for you.

13. $9^2 = 81$

$(9^2 = 9 \times 9 = 81)$

14. $3^2 =$ ____   15. $10^2 =$ ____

16. What is the GCF of 21 and 24?

17. Is 123 divisible by three?

18. Find the prime factors of 50.

19. Erin ate $\frac{1}{2}$ of a blueberry pie and $\frac{2}{3}$ of an apple pie. How much pie did she eat? Give your answer as a mixed number.

20. Jessica has to make a trip of 8,925 miles. If she travels 425 miles a day, how long will the trip take?

In the Application and Enrichment pages for lesson 3, you learned how to use parentheses to show what part of a problem should be completed first.

Example 1

$6 \times (3 + 5) =$

Compute the numbers inside the parentheses first and then finish the problem.

$6 \times (3 + 5) = 6 \times (8) = 48$

Some problems are more complex and use brackets. Brackets are a different style of parentheses. They are used to help prevent confusion in problems with several steps. Compute the part inside the parentheses first and then calculate the part inside the brackets. Finally, perform any other operations to finish the problem.

Example 2

$[5 \times (2 + 4)] - 3 =$

First compute the numbers inside the parentheses.

$[5 \times (6)] - 3 =$

Next calculate the numbers inside the brackets.

$[30] - 3 =$

Subtract to finish solving the problem. You may remove the parentheses and brackets as soon as they are no longer needed.

$30 - 3 = 27$

Solve each problem by working inside the parentheses first and then inside the brackets. Finish the problem by completing any other required operations.

1. $[(24 \div 3) + 1] \times 5 =$

2. $[(16 - 6) \times 9] + 2 =$

3. $2 \times [(3 + 4) - 5] =$

4. $7 + 4[9 - (2 \times 3)] =$

These mental math problems are from lesson 15 in the instruction manual. Solve each problem mentally. Then rewrite the problem with numerals, using parentheses and brackets as needed. Solve the new expression and compare your answers.

5. Twenty-four divided by four, plus two, times six, equals?

6. Fifteen minus eight, times two, plus five, equals?

7. Three plus three, times five, divided by 10, equals?

At times it may be useful to convert a fraction with a denominator of 10 to a fraction with a denominator of 100. Study the examples to see how to apply your math skills to do this.

Example 2
Simplify the fraction 80/100.

$$\frac{80}{100} \div \frac{10}{10} = \frac{8}{10}$$

Example 3
Rewrite 7/10 as an equivalent fraction with a denominator of 100.

$$\frac{7}{10} \times \frac{10}{10} = \frac{70}{100}$$

Convert the fractions with a denominator of 10 to fractions with a denominator of 100. Then add or subtract the fractions.

8. $\dfrac{2}{10} + \dfrac{70}{100} =$

9. $\dfrac{4}{10} - \dfrac{1}{100} =$

10. $\dfrac{10}{100} + \dfrac{2}{10} =$

Write the length of each line. Always write the fractional part of your answer in simplified form. The drawings on these pages are a slightly longer than an inch to make them easier to read. The first one has been done for you.

1. $3\frac{1}{2}$ in

2. _____ in

3. _____ in

4. _____ in

5. _____ in

6. _____ in

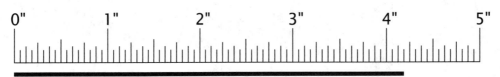

16B

Write the length of each line. Always write the fractional part of your answer in simplified form.

1. _____ in

2. _____ in

3. _____ in

4. _____ in

5. _____ in

6. _____ in

16C

Write the length of each line. Always write the fractional part of your answer in simplified form.

1. ____ in

2. ____ in

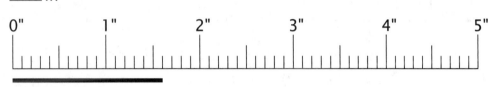

3. ____ in

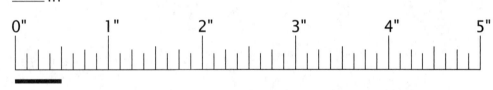

4. ____ in

5. _____ in

6. _____ in

Write the length of the line.

1. _____ in

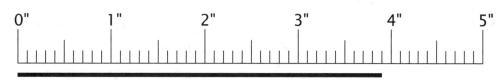

Change each mixed number to an improper fraction.

2. $2\dfrac{6}{8} = \dfrac{}{8} + \dfrac{}{8} + \dfrac{}{8} =$ _____

3. $3\dfrac{1}{6} = \dfrac{}{6} + \dfrac{}{6} + \dfrac{}{6} + \dfrac{}{6} =$ _____

Change each improper fraction to a mixed number.

4. $\dfrac{7}{5} = \dfrac{}{5} + \dfrac{}{5} =$ _____

5. $\dfrac{19}{8} = \dfrac{}{8} + \dfrac{}{8} + \dfrac{}{8} =$ _____

Simplify the fractions.

6. $\dfrac{15}{30} =$ _____

7. $\dfrac{28}{70} =$ _____

8. $\dfrac{6}{9} =$ _____

Add or subtract. Write your answer to #11 as a mixed number.

9. $\dfrac{3}{8} + \dfrac{2}{5} =$ _____

10. $\dfrac{5}{7} - \dfrac{1}{2} =$ _____

11. $\dfrac{1}{2} + \dfrac{3}{4} + \dfrac{7}{8} =$ _____ = _____

QUICK REVIEW

A triangle is one half of a rectangle. To find the area of a triangle, multiply the base by the height and find one half of the answer (divide by two). The height is a line that makes a right angle (or square corner) with the base.

Find the area of each triangle. The first one has been done for you.

12. 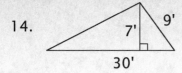   ($\frac{1}{2}$)(12 ft)(8 ft) = 48 sq ft   A = _48 sq ft_

13. 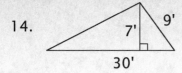   A = _____

14. 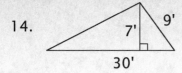   A = _____

15. Find the prime factors of 75.

16. What is the area of a square room that measures 15 feet on each side?

17. One half of the house has been painted. Jeff did $\frac{4}{10}$ of what has been done. What part of the house did Jeff paint? Simplify your answer.

18. John is 3,279 days old. How many weeks old is he?

Write the length of the line.

1. ____ in

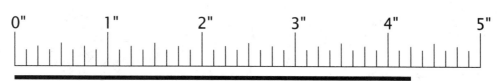

Change each mixed number to an improper fraction.

2. $2\dfrac{1}{10} = \dfrac{}{10} + \dfrac{}{10} + \dfrac{}{10} = \underline{}$

3. $3\dfrac{2}{3} = \dfrac{}{3} + \dfrac{}{3} + \dfrac{}{3} + \dfrac{}{3} = \underline{}$

Change each improper fraction to a mixed number.

4. $\dfrac{17}{9} = \dfrac{}{9} + \dfrac{}{9} = \underline{}$

5. $\dfrac{5}{2} = \dfrac{}{2} + \dfrac{}{2} + \dfrac{}{2} = \underline{}$

Divide. Write all your answers as mixed numbers. Be sure to simplify the fractions if possible.

6. $\dfrac{5}{8} \div \dfrac{1}{4} = \underline{}$ 7. $\dfrac{5}{7} \div \dfrac{1}{2} = \underline{}$

8. $\dfrac{3}{4} \div \dfrac{1}{2} = \underline{}$

Multiply the numerators and denominators by the same numbers to make equivalent fractions.

9. $\dfrac{1}{5} = \dfrac{2}{} = \dfrac{}{15}$

10. $\dfrac{3}{4} = \dfrac{}{} = \dfrac{9}{}$

11. $\dfrac{4}{7} = \dfrac{}{} = \dfrac{12}{}$

Find the area. Include a fraction in the answer if you cannot divide evenly.

12. 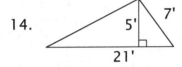   A = _____

13. A = _____

14. A = _____

15. What is the GCF of 15 and 20?

16. What number is equivalent to 15^2?

17. What is the perimeter of a triangle that measures $\frac{1}{2}$ inch on each side? Write your answer as a mixed number.

18. A farmer has a rectangular piece of land that measures 256 yards by 38 yards. What is the area of the piece of land?

19. Karen spent $\frac{3}{15}$ of her income on books. Simplify the fraction and tell what part of her income was spent on books.

20. Dennis and Danny have equal sections of lawn to mow. Dennis has mowed $\frac{5}{6}$ of his section, and Danny has finished $\frac{11}{12}$ of his section. Which one has mowed more lawn so far?

Write the length of the line.

1. _____ in

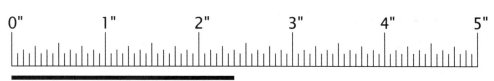

Change each mixed number to an improper fraction.

2. $2\dfrac{3}{4} = \dfrac{}{4} + \dfrac{}{4} + \dfrac{}{4} = \underline{}$

3. $3\dfrac{5}{7} = \dfrac{}{7} + \dfrac{}{7} + \dfrac{}{7} + \dfrac{}{7} = \underline{}$

Change each improper fraction to a mixed number.

4. $\dfrac{11}{6} = \dfrac{}{6} + \dfrac{}{6} = \underline{}$

5. $\dfrac{27}{10} = \dfrac{}{10} + \dfrac{}{10} + \dfrac{}{10} = \underline{}$

Multiply. Simplify if possible.

6. $\dfrac{1}{5} \times \dfrac{5}{6}$

7. $\dfrac{2}{3} \times \dfrac{4}{5}$

8. $\dfrac{1}{3} \times \dfrac{2}{8}$

Use the Rule of Four to make denominators the same and then compare the fractions.

9. $\dfrac{2}{3} \bigcirc \dfrac{3}{4}$

10. $\dfrac{3}{8} \bigcirc \dfrac{1}{7}$

11. $\frac{5}{9}$ 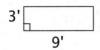 $\frac{4}{11}$

Find the area and perimeter of each figure.

3' ☐
9'

12. A = _____   13. P = _____

17" ☐
17"

14. A = _____   15. P = _____

12' 20'
16'

16. A = _____   17. P = _____

18. Find the prime factors of 100.

19. What number is equivalent to 20^2?

20. A piece of yarn is $\frac{5}{8}$ of a yard long. Sue wants to cut the yarn into equal pieces that are each $\frac{1}{16}$ of a yard long. How many pieces will she have?

The Smith family has several members. Some are male, and some are female. Using a different way of classifying the members, you can say that some are parents and some are children. Using eye color, you get other groups. No matter what classification you use, everyone is still a member of the Smith family.

Two-dimensional shapes are described by their sides and by their angles. Shapes can also belong to more than one group. For example, squares are *quadrilaterals* because they have four sides. Since they have two pairs of parallel sides, they fit in the special group of quadrilaterals called *parallelograms.* Parallelograms with four right angles are *rectangles,* and rectangles with all four sides the same length are *squares.* There can be many ways to describe one shape.

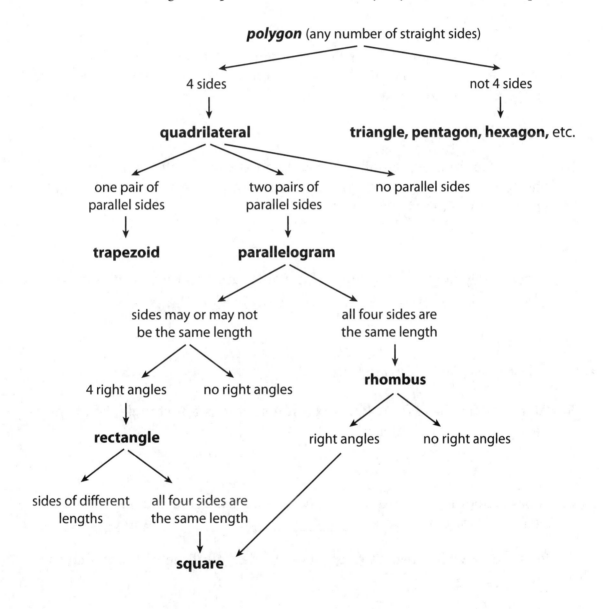

Use words from the box to name each drawing described below. Try to use all the names that apply to each shape.

parallelogram	quadrilateral	rectangle
rhombus	square	trapezoid

1. Eva drew a polygon. It had four sides. It had only one pair of parallel sides. What shape did she draw?

2. Timothy drew a polygon that has two pairs of parallel sides. All four sides were the same length, but the polygon didn't have any right angles. What shape did Timothy draw?

3. Carrie also drew a polygon that had two pairs of parallel sides. Her shape had four right angles, but the sides were not all the same length. What shape did she draw?

4. Becca's polygon had four sides of equal length and four right angles. What shape did Becca draw?

Fill in the blanks to make the statements true.

5. All rectangles have four right angles. A square is a rectangle, so a square has _____ right angles.

6. A parallelogram has two pairs of parallel sides. Some shapes that have two pairs of parallel sides are _____, _____, and _____

   Because they have two pairs of parallel sides, all three of those shapes are _____.

Estimate first. Then add the mixed numbers. When rounding mixed numbers, use your knowledge of equivalent fractions to decide if the fraction is greater than or less than one half. If the fraction is less than one half, round to the given whole number. If the fraction is one half or greater, round up to the next whole number. The first one has been done for you.

1. $3 \frac{4}{7}$ （4）

    $+ \ 5 \frac{1}{7}$ （5）

    $8 \frac{5}{7}$ （≈ 9）

2. $3 \frac{1}{6}$ ◯

    $+ \ 2 \frac{4}{6}$ ◯

    （≈）

3. $7 \frac{2}{5}$ ◯

    $+ \ 9 \frac{2}{5}$ ◯

    （≈）

4. $2 \frac{3}{9}$ ◯

    $+ \ 1 \frac{5}{9}$ ◯

    （≈）

5. $4 \frac{2}{7}$ ◯

    $+ \ 8 \frac{4}{7}$ ◯

    （≈）

6. $6 \frac{2}{4}$ ◯

    $+ \ 7 \frac{1}{4}$ ◯

    （≈）

Estimate first and then subtract the mixed numbers. The first one has been done for you.

7.
$$7 \frac{2}{3} \quad \boxed{8}$$
$$- \ 5 \frac{1}{3} \quad \boxed{5}$$
$$2 \frac{1}{3} \quad \boxed{\approx 3}$$

8.
$$6 \frac{4}{8} \quad \bigcirc$$
$$- \ 3 \frac{1}{8} \quad \bigcirc$$
$$\approx \bigcirc$$

9.
$$10 \frac{4}{5} \quad \bigcirc$$
$$- \ 9 \frac{3}{5} \quad \bigcirc$$
$$\approx \bigcirc$$

10.
$$8 \frac{7}{9} \quad \bigcirc$$
$$- \ 4 \frac{5}{9} \quad \bigcirc$$
$$\approx \bigcirc$$

11.
$$5 \frac{3}{4} \quad \bigcirc$$
$$- \ 3 \frac{2}{4} \quad \bigcirc$$
$$\approx \bigcirc$$

12.
$$3 \frac{5}{6} \quad \bigcirc$$
$$- \ 2 \frac{4}{6} \quad \bigcirc$$
$$\approx \bigcirc$$

LESSON PRACTICE

Estimate and then add.

1. $9\ \frac{3}{6}$

    $+\ 9\ \frac{2}{6}$

    $\approx$

2. $8\ \frac{2}{8}$

    $+\ 3\ \frac{5}{8}$

    $\approx$

3. $7\ \frac{1}{3}$

    $+\ 4\ \frac{1}{3}$

    $\approx$

4. $6\ \frac{1}{4}$

    $+\ 3\ \frac{2}{4}$

    $\approx$

5. $2\ \frac{1}{5}$

    $+\ 4\ \frac{3}{5}$

    $\approx$

6. $9\ \frac{4}{7}$

    $+\ 4\ \frac{2}{7}$

    $\approx$

Estimate and then subtract.

7. $2\frac{4}{5}$ ◯
   $-\ 1\frac{2}{5}$ ◯
   _____ $\approx$ ◯

8. $8\frac{5}{8}$ ◯
   $-\ 6\frac{2}{8}$ ◯
   _____ $\approx$ ◯

9. $5\frac{3}{5}$ ◯
   $-\ 4\frac{2}{5}$ ◯
   _____ $\approx$ ◯

10. $4\frac{6}{7}$ ◯
    $-\ 3\frac{2}{7}$ ◯
    _____ $\approx$ ◯

11. $9\frac{5}{6}$ ◯
    $-\ 7\frac{4}{6}$ ◯
    _____ $\approx$ ◯

12. $16\frac{5}{9}$ ◯
    $-\ 8\frac{1}{9}$ ◯
    _____ $\approx$ ◯

13. Emily had $2\frac{1}{3}$ apple pies and $1\frac{1}{3}$ pumpkin pies on the day after Thanksgiving. How much pie did she have in all?

14. Nick bought $4\frac{3}{5}$ of a pound of peppermints. He gave $3\frac{1}{5}$ pounds of them to Sean. How many pounds of peppermints does he have left?

Estimate and then add.

1. $5 \frac{3}{8}$ ⭕

     $+ 7 \frac{4}{8}$ ⭕

       $\approx$

2. $3 \frac{7}{9}$ ⭕

     $+ 4 \frac{1}{9}$ ⭕

       $\approx$

3. $8 \frac{1}{4}$ ⭕

     $+ 5 \frac{2}{4}$ ⭕

       $\approx$

4. $2 \frac{3}{10}$ ⭕

     $+ 6 \frac{4}{10}$ ⭕

       $\approx$

5. $9 \frac{1}{5}$ ⭕

     $+ 1 \frac{2}{5}$ ⭕

       $\approx$

6. $6 \frac{2}{9}$ ⭕

     $+ 3 \frac{2}{9}$ ⭕

       $\approx$

LESSON PRACTICE 17C

Estimate and then subtract.

7. $12\frac{4}{7}$ ◯
    $-\ 4\frac{2}{7}$ ◯
    ⟨≈⟩

8. $8\frac{2}{8}$ ◯
    $-\ 5\frac{1}{8}$ ◯
    ⟨≈⟩

9. $14\frac{3}{5}$ ◯
    $-\ 9\frac{1}{5}$ ◯
    ⟨≈⟩

10. $2\frac{5}{11}$ ◯
    $-\ 1\frac{3}{11}$ ◯
    ⟨≈⟩

11. $7\frac{2}{3}$ ◯
    $-\ 3\frac{1}{3}$ ◯
    ⟨≈⟩

12. $15\frac{7}{8}$ ◯
    $-\ 9\frac{2}{8}$ ◯
    ⟨≈⟩

13. Ryan counted his dimes and discovered that he had $4\frac{7}{10}$ dollars. He spent $1\frac{4}{10}$ dollars. How many dollars does he have left?

14. Madison ran $1\frac{1}{8}$ miles in the morning and $1\frac{2}{8}$ miles in the afternoon. How far did she run in all?

Add or subtract. Continue to estimate your answers mentally.

1.
$$\begin{array}{r} 3\frac{3}{7} \\ + 2\frac{1}{7} \\ \hline \end{array}$$

2.
$$\begin{array}{r} 11\frac{3}{4} \\ - 8\frac{2}{4} \\ \hline \end{array}$$

3.
$$\begin{array}{r} 7\frac{1}{5} \\ + 2\frac{3}{5} \\ \hline \end{array}$$

Change each mixed number to an improper fraction. Write out the steps only if you need to.

4. $3\frac{4}{5} = $ _____

5. $4\frac{1}{8} = $ _____

6. $5\frac{5}{6} = $ _____

Change each improper fraction to a mixed number. These are also division problems.

7. $\frac{19}{9} = $ _____

8. $\frac{24}{5} = $ _____

9. $\frac{15}{8} = $ _____

QUICK REVIEW

A cube is a three-dimensional shape. It has six faces that are squares and twelve edges of equal length. Volume measures the space inside the shape and is expressed in cubic units. To find the volume of a cube, first find the area of the base of the cube and multiply by the height.

Find the volume of each cube. The first one has been done for you.

10. 3" 3" 3" $(3 \text{ in} \times 3 \text{ in})(3 \text{ in}) = 27 \text{ in}^3$ V = <u>27 cubic inches (in^3)</u>

11. 20' 20' 20'   V = _____

12. 12' 12' 12'   V = _____

13. The floor of Hailey's room measures 10 ft by 10 ft. The ceiling is also 10 ft high. How many cubic feet of air does her room hold?

14. It is $7\frac{2}{3}$ miles from my house to Bethany's house and $3\frac{1}{3}$ miles from my house to Pat's house. How much farther do I have to travel to get to Bethany's house than to get to Pat's house?

15. Find the prime factors of 150.

16. What number is equivalent to 4^2?

17. Andy has a rectangular garden that measures 9 feet by 10 feet. He bought a bag of fertilizer that was enough for 100 square feet. Does Andy have enough fertilizer for his garden?

18. Eighteen birds landed in Tom's yard. One sixth of them were robins. How many robins were there?

Add or subtract. Continue to estimate your answers mentally.

1. $2 \dfrac{4}{5}$

    $- 1 \dfrac{2}{5}$

2. $4 \dfrac{1}{9}$

    $+ 6 \dfrac{7}{9}$

3. $5 \dfrac{5}{6}$

    $- 2 \dfrac{4}{6}$

Change each mixed number to an improper fraction. Write out the steps only if you need to.

4. $2 \dfrac{1}{2} =$ _____

5. $6 \dfrac{2}{4} =$ _____

6. $5 \dfrac{3}{8} =$ _____

Change each improper fraction to a mixed number.

7. $\dfrac{51}{10} =$ _____

8. $\dfrac{55}{9} =$ _____

9. $\dfrac{16}{5} =$ _____

Write the length of the line.

10. _____ in

Find the volume of each cube.

11. 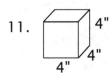 4" 4" 4"

V = _____

12. 15' 15' 15'

V = _____

13. 30' 30' 30'

V = _____

14. A cube-shaped container measures one foot on each edge. How many cubic feet of water will the container hold?

15. What is the GCF of 25 and 30?

16. Mr. King harvested $4\frac{1}{4}$ tons of hay from one field and $3\frac{2}{4}$ tons from another field. How many tons of hay has he harvested in all?

17. The three sides of a triangle are $\frac{1}{3}$ of a yard, $\frac{2}{3}$ of a yard, and $\frac{3}{5}$ of a yard. What is the perimeter of the triangle? Simplify your answer and write it as a mixed number.

18. Kaitlyn had been assigned $\frac{4}{5}$ of the job. She has completed $\frac{1}{2}$ of what she was assigned. What part of the total job has Kaitlyn completed?

Add or subtract. Continue to estimate your answers mentally.

1. $4\dfrac{2}{5}$

   $+\ 8\dfrac{2}{5}$

   _____

2. $21\dfrac{3}{7}$

   $-\ 9\dfrac{1}{7}$

   _____

3. $6\dfrac{1}{8}$

   $+\ 7\dfrac{4}{8}$

   _____

Change each mixed number to an improper fraction. Write out the steps only if you need to.

4. $1\dfrac{3}{4} =$ _____

5. $2\dfrac{5}{6} =$ _____

6. $6\dfrac{1}{3} =$ _____

Change each improper fraction to a mixed number.

7. $\dfrac{5}{4} =$ _____

8. $\dfrac{23}{6} =$ _____

9. $\dfrac{11}{4} =$ _____

Follow the signs. Simplify if possible.

10. $\dfrac{1}{2} \times \dfrac{3}{5} =$ _____

11. $\dfrac{2}{3} \div \dfrac{1}{6} =$ _____

12. $\dfrac{4}{7} \times \dfrac{1}{4} =$ _____

Find the volume of each cube.

13. 7" 7" 7" V = _____

14. 25' 25' 25' V = _____

15. 41' 41' 41' V = _____

16. Daniel's alphabet blocks are cubes that measure two inches along each side. What is the volume of one of his blocks? What is the volume of six blocks?

17. Is 234 divisible by 9?

18. Hannah bought $2\frac{1}{5}$ pounds of apples and $3\frac{2}{5}$ pounds of bananas. How many pounds of fruit did she buy?

19. A rectangle is $\frac{1}{2}$ of a mile long and $\frac{1}{4}$ of a mile wide. What is the area of the rectangle? Solve this in the same way as you would solve a whole number area problem. Your answer will be a fraction of a square mile.

20. Bryan has two bags of jelly beans. One weighs $\frac{3}{7}$ of a pound, and the other weighs $\frac{4}{11}$ of a pound. Which bag is heavier?

Sometimes you will come across area problems where one or both sides of the figure are fractional lengths. Drawing these types of problems can make the results clearer.

Example 1
The square represents one square mile (1 mi × 1 mi = 1 sq mi).

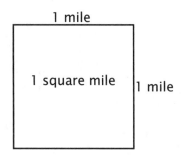

Find the area of a piece of land that measures one half of a mile on each side. Multiply to find the area. $\frac{1}{2}$ mile × $\frac{1}{2}$ mile = $\frac{1}{4}$ square mile

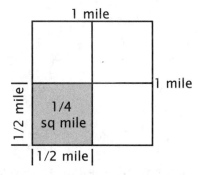

The drawing helps us see that the area of a plot that is one half of a mile on each side has an area that is one fourth, or one quarter, of a plot that is one mile on each side.

Example 2
What is the area of a plot that measures one third of a mile by two miles?

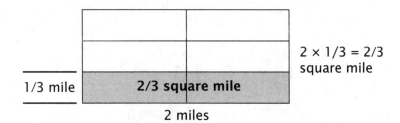

The drawing shows each $\frac{1}{3}$ part of the area in a different square. Visualize the shaded part on the right moved to an empty space in the left-hand square. You can see that the answer is $\frac{2}{3}$ of one.

Now it is your turn. Read each area problem and make a drawing similar to those on the previous page. Shade the part of the drawing that represents the answer. Your drawing does not have to be to scale, but it should give a good idea of why the answer is true.

1. What is the area of a poster that measures one third of a yard on each side?

2. What is the area of a plot of land that measures one half of a mile on one side and one third of a mile on the other side?

3. What is the area of a garden that measures two yards by one half of a yard? Just as in Example 2 on the previous page, you may want to move one shaded part to another position to represent the answer more clearly. You can do this mentally or make a second drawing.

Add the mixed numbers. Regroup as necessary so that your answers do not include improper fractions. Simplify if possible. The first two have been done for you.

1. $2 \frac{5}{8}$

    $+ 1 \frac{7}{8}$

    $3 \frac{12}{8}$

    $3 + 1 \frac{4}{8} = 4 \frac{4}{8} = 4 \frac{1}{2}$

2. $1 \frac{2}{3}$

    $+ 1 \frac{1}{3}$

    $2 \frac{3}{3}$

    $2 + 1 = 3$

3. $1 \frac{2}{5}$

    $+ 4 \frac{4}{5}$

4. $3 \frac{3}{8}$

    $+ 1 \frac{7}{8}$

5. $7 \frac{2}{5}$

    $+ 2 \frac{4}{5}$

6. $2 \frac{5}{6}$

    $+ 4 \frac{3}{6}$

7. $\quad 4\ \dfrac{2}{4}$

$\quad +\ 2\ \dfrac{3}{4}$

8. $\quad 5\ \dfrac{5}{8}$

$\quad +\ 7\ \dfrac{3}{8}$

9. $\quad 2\ \dfrac{3}{7}$

$\quad +\ 5\ \dfrac{6}{7}$

10. $\quad 3\ \dfrac{1}{2}$

$\quad +\ 3\ \dfrac{1}{2}$

11. $\quad 7\ \dfrac{2}{3}$

$\quad +\ 3\ \dfrac{2}{3}$

12. $\quad 4\ \dfrac{8}{9}$

$\quad +\ 8\ \dfrac{5}{9}$

Add the mixed numbers. Regroup as necessary so your answers do not include improper fractions. Simplify if possible.

1. $9\dfrac{4}{6}$
   $+\ 7\dfrac{5}{6}$

2. $3\dfrac{4}{5}$
   $+\ 6\dfrac{3}{5}$

3. $2\dfrac{7}{8}$
   $+\ 5\dfrac{7}{8}$

4. $4\dfrac{3}{4}$
   $+\ 1\dfrac{1}{4}$

5. $2\dfrac{4}{9}$
   $+\ 3\dfrac{6}{9}$

6. $6\dfrac{2}{3}$
   $+\ 8\dfrac{2}{3}$

7. $1 \dfrac{5}{11}$

    $+ \ 2 \dfrac{6}{11}$
    _____

8. $7 \dfrac{7}{10}$

    $+ \ 9 \dfrac{5}{10}$
    _____

9. $2 \dfrac{4}{7}$

    $+ \ 2 \dfrac{6}{7}$
    _____

10. $5 \dfrac{1}{3}$

    $+ \ 2 \dfrac{2}{3}$
    _____

11. $6 \dfrac{4}{5}$

    $+ \ 4 \dfrac{4}{5}$
    _____

12. $10 \dfrac{5}{8}$

    $+ \ 3 \dfrac{7}{8}$
    _____

Add the mixed numbers. Regroup as necessary so your answers do not include improper fractions. Simplify if possible.

1. $4 \frac{5}{8}$

    $+ 3 \frac{3}{8}$

2. $4 \frac{5}{6}$

    $+ 7 \frac{4}{6}$

3. $3 \frac{8}{9}$

    $+ 6 \frac{7}{9}$

4. $5 \frac{4}{5}$

    $+ 2 \frac{2}{5}$

5. $3 \frac{5}{10}$

    $+ 4 \frac{7}{10}$

6. $7 \frac{3}{4}$

    $+ 9 \frac{2}{4}$

7. $2\dfrac{6}{12}$

    $+\ 3\dfrac{7}{12}$

8. $8\dfrac{8}{11}$

    $+10\dfrac{6}{11}$

9. $6\dfrac{5}{8}$

    $+\ 5\dfrac{7}{8}$

10. $6\dfrac{2}{4}$

    $+\ 4\dfrac{2}{4}$

11. $7\dfrac{5}{6}$

    $+\ 5\dfrac{4}{6}$

12. $12\dfrac{5}{9}$

    $+\ 8\dfrac{8}{9}$

From now on simplify all fractions unless the directions state otherwise. Improper fractions should be written as mixed numbers.

Add the mixed numbers.

1. $\quad 4 \dfrac{3}{7}$

    $+ \ 3 \ \dfrac{6}{7}$
    _____

2. $\quad 14 \ \dfrac{4}{5}$

    $+ \ 9 \ \dfrac{2}{5}$
    _____

3. $\quad 8 \ \dfrac{7}{9}$

    $+ \ 5 \ \dfrac{4}{9}$
    _____

Change each mixed number to an improper fraction.

4. $5 \dfrac{1}{3} = $ _____

5. $7 \dfrac{2}{9} = $ _____

6. $3 \dfrac{4}{5} = $ _____

Change each improper fraction to a mixed number.

7. $\dfrac{27}{5} = $ _____

8. $\dfrac{39}{6} = $ _____

9. $\dfrac{33}{10} = $ _____

QUICK REVIEW

A rectangular solid has six faces that are rectangles. To calculate the volume of a rectangular solid, first find the area of the base of the solid and then multiply by the height. Answers should be given in cubic units.

Find the volume of each rectangular solid. The first one has been done for you.

10.

$(3 \text{ in} \times 4 \text{ in})(2 \text{ in}) = 24 \text{ in}^3$ V = <u>24 cu in</u>

11.

V = _____

12.

V = _____

13. The floor of Brian's room measures 12 feet by 13 feet, and the flat ceiling is 8 feet high. How many cubic feet of air does his room hold?

14. At birth, one of a set of twins weighed $4\frac{5}{8}$ pounds, and the other weighed $3\frac{7}{8}$ pounds. What is the combined weight of the twins?

15. Find the prime factors of 74.

16. What number is equivalent to 16^2?

17. A triangle has a base of 10 feet and a height of 15 feet. What is the area of the triangle?

18. A group of seven students took a road trip. The trip covered 3,885 miles, and they shared the driving equally. How many miles did each one drive?

SYSTEMATIC REVIEW

Add the mixed numbers.

1. $\begin{array}{r} 2\dfrac{4}{5} \\ +\ 1\dfrac{2}{5} \\ \hline \end{array}$

2. $\begin{array}{r} 4\dfrac{1}{9} \\ +\ 6\dfrac{7}{9} \\ \hline \end{array}$

3. $\begin{array}{r} 5\dfrac{5}{6} \\ +\ 2\dfrac{4}{6} \\ \hline \end{array}$

Follow the signs.

4. $\dfrac{5}{6} - \dfrac{1}{3} = \underline{\hspace{1cm}}$

5. $\dfrac{2}{3} - \dfrac{1}{8} = \underline{\hspace{1cm}}$

6. $\dfrac{3}{4} + \dfrac{4}{5} = \underline{\hspace{1cm}}$

7. $\dfrac{5}{7} \times \dfrac{1}{2} = \underline{\hspace{1cm}}$

8. $\dfrac{3}{5} \div \dfrac{3}{4} = \underline{\hspace{1cm}}$

9. $\dfrac{1}{4} \times \dfrac{5}{6} = \underline{\hspace{1cm}}$

Write the length of the line.

10.   in

Find the volume of each rectangular solid.

11.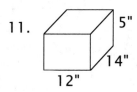

12.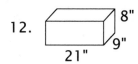

13.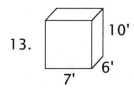

14. A rectangular swimming pool is 20 feet long, 15 feet wide, and 6 feet deep. How many cubic feet of water will it hold if it is filled to the top?

15. A cubic foot of water weighs about 63 pounds. How much will the water in #14 weigh when the pool is filled?

16. What is the GCF of 16 and 20?

17. One side of a square is 32 feet. How much fence is needed to fence one half of the perimeter?

18. Kevin was $5\frac{2}{3}$ feet tall on his last birthday. This year he grew $\frac{1}{3}$ of a foot. How tall is he now?

SYSTEMATIC REVIEW

Add the mixed numbers.

1. $5 \frac{4}{6}$
   $+ 9 \frac{2}{6}$

2. $11 \frac{5}{8}$
   $+ 3 \frac{7}{8}$

3. $7 \frac{7}{10}$
   $+ 8 \frac{9}{10}$

Follow the signs.

4. $\frac{5}{7} - \frac{2}{9} =$ _____

5. $\frac{1}{8} + \frac{4}{5} =$ _____

6. $\frac{6}{7} - \frac{1}{8} =$ _____

7. $\frac{5}{6} \div \frac{7}{8} =$ _____

8. $\frac{1}{2} \times \frac{3}{5} =$ _____

9. $\frac{2}{3} \div \frac{1}{6} =$ _____

Solve. Remember that the multiplication sign means the same as "of."

10. $\frac{4}{7} \times 56 =$ _____

11. $\frac{3}{5} \times 50 =$ _____

12. $\frac{1}{2} \times 14 =$ _____

Find the volume of each rectangular solid.

13. 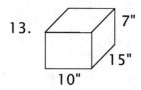 V = _____

14. 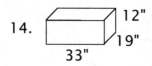 V = _____

15. 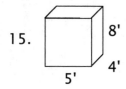 V = _____

16. A rectangular swimming pool is 25 feet long, 13 feet wide, and 8 feet deep. How many cubic feet of water will it hold if it is filled to half of its depth?

17. At 63 pounds per cubic foot, how much will the water in #16 weigh?

18. Is 205 divisible by 5?

19. A rectangle is $\frac{1}{2}$ of a mile long and $\frac{1}{4}$ of a mile wide. What is the perimeter of the rectangle? Add two fractions at a time and then add the two sums.

20. Ashley has $\frac{4}{5}$ of a pie that she wants to divide evenly among four people. How much pie should each person receive?

We can find the volume of a cube or a rectangular solid by multiplying the length by the width by the height. Build some three-dimensional shapes with the Math-U-See blocks or other cubes to help you see why this is true.

Example 1

Use six cubes to build a rectangle that is three cubes by two cubes.

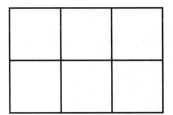

This is the top view of your six cubes. The area of the rectangle is 3 units × 2 units, or 6 square units.

Now stack another six cubes on top of the first layer.

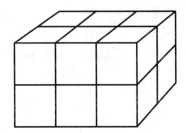

Multiply the length by the width by the height. The result is 3 units × 2 units × 2 units, or 12 cubic units.

Take apart the rectangular solid that you just built and count the blocks or cubes. You should have 12 of them.

Rebuild your "house" and add another story to it. Now what is the volume of the entire building?

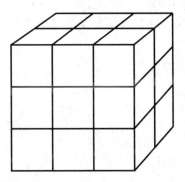

The area of the base of the building is 3 units × 2 units, or 6 square units, as before. Since we added another story to the building, the volume is now area of the base × 3 units. The volume is 6 square units × 3 units, or 18 cubic units.

You can find the volume of three-dimensional shapes that are made up of more than one rectangular solid by adding the volumes of each part.

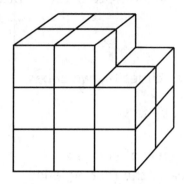

The volume of the bottom structure (first two layers) is 12 cubic units.

The top layer is 2 units × 2 units × 1 unit, or 4 cubic units. 12 cubic units + 4 cubic units = 16 cubic units

Count the blocks to see if the answer you computed is correct.

Find the area of each structure by computing the volume of the rectangular solids. If you have enough blocks or cubes, find the area by building and counting as well. A box of sugar cubes may be an inexpensive source for solids that have a greater number of cubes.

1. The base of a tower is 3 yards by 4 yards. The height of the main part of the tower is 6 yards. At one corner of the tower is a smaller tower that is 1 yard by 1 yard by 3 yards. What is the volume of the entire tower?

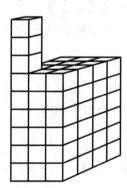

2. The floor plan of a small building looks like this. Each square is one yard on a side. The height of the entire building is five yards. What is the total volume of the building?

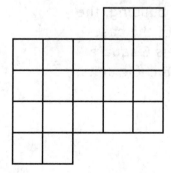

Regroup and subtract the mixed numbers. The first two have been done for you.

1.
$$\overset{4}{\cancel{5}}\ \frac{1}{4} \quad\overset{\frown}{\frac{4}{4}}\quad 4\ \frac{5}{4}$$
$$-\ 2\ \frac{3}{4} \qquad\qquad -\ 2\ \frac{3}{4}$$
$$\qquad\qquad\qquad\qquad 2\ \frac{2}{4}\ =\ 2\ \frac{1}{2}$$

2.
$$\overset{2}{\cancel{3}} \quad\overset{\frown}{\frac{3}{3}}\quad 2\ \frac{3}{3}$$
$$-\ 1\ \frac{2}{3} \qquad\qquad -\ 1\ \frac{2}{3}$$
$$\qquad\qquad\qquad\qquad 1\ \frac{1}{3}$$

3.
$$6\ \frac{1}{5}$$
$$-\ 2\ \frac{3}{5}$$

4.
$$4\ \frac{1}{8}$$
$$-\ 1\ \frac{5}{8}$$

5.
$$2\ \frac{1}{4}$$
$$-\ 1\ \frac{2}{4}$$

6.
$$6$$
$$-\ 3\ \frac{7}{8}$$

7. $7 \frac{5}{16}$

     $- 2 \frac{7}{16}$

8. $8 \frac{3}{5}$

     $- 2 \frac{3}{5}$

9. $4 \frac{1}{6}$

     $- 1 \frac{5}{6}$

10. $9 \frac{1}{8}$

     $- 2 \frac{7}{8}$

11. 4

     $- \frac{1}{3}$

12. $7 \frac{2}{5}$

     $- 6 \frac{4}{5}$

Regroup and subtract the mixed numbers.

1. $14 \dfrac{3}{7}$
    $- \ 2 \dfrac{4}{7}$
    _____

2. 11
    $- \ 2 \dfrac{8}{9}$
    _____

3. $8 \dfrac{3}{10}$
    $- \ 5 \dfrac{4}{10}$
    _____

4. $19 \dfrac{2}{6}$
    $- 10 \dfrac{3}{6}$
    _____

5. $3 \dfrac{1}{3}$
    $- \ 1 \dfrac{2}{3}$
    _____

6. 5
    $- \ 1 \dfrac{4}{7}$
    _____

7. $5 \dfrac{1}{4}$
$- \ 2 \dfrac{3}{4}$
_____

8. $6 \dfrac{1}{5}$
$- \ 2 \dfrac{3}{5}$
_____

9. $4 \dfrac{1}{8}$
$- \ 1 \dfrac{5}{8}$
_____

10. $6 \dfrac{5}{8}$
$- \ 3 \dfrac{7}{8}$
_____

11. 8
$- \ \dfrac{3}{10}$
_____

12. $2 \dfrac{1}{4}$
$- \ 1 \dfrac{3}{4}$
_____

LESSON PRACTICE

Regroup and subtract the mixed numbers.

1. $5 \frac{1}{3}$

   $- 2 \frac{2}{3}$

   _____

2. 6

   $- 2 \frac{3}{8}$

   _____

3. $6 \frac{3}{16}$

   $- 2 \frac{5}{16}$

   _____

4. $8 \frac{1}{5}$

   $- 2 \frac{4}{5}$

   _____

5. $9 \frac{1}{4}$

   $- 6 \frac{3}{4}$

   _____

6. 9

   $- 1 \frac{3}{5}$

   _____

7. $4 \frac{1}{6}$

    $- \ 1 \frac{1}{6}$
_____

8. $5 \frac{5}{8}$

    $- \ 2 \frac{7}{8}$
_____

9. $6 \frac{1}{3}$

    $- \ 2 \frac{2}{3}$
_____

10. $18 \frac{1}{4}$

    $- \ 7 \frac{2}{4}$
_____

11. 6

    $- \ \frac{4}{5}$
_____

12. $9 \frac{1}{6}$

    $- \ 2 \frac{5}{6}$
_____

SYSTEMATIC REVIEW

Add or subtract the mixed numbers. Regroup as needed.

1. $5\frac{1}{5}$
$-\,2\frac{3}{5}$

2. $8\frac{5}{8}$
$-\,2\frac{7}{8}$

3. 6
$-\,4\frac{2}{3}$

4. $7\frac{1}{5}$
$+\,2\frac{4}{5}$

5. $1\frac{3}{4}$
$+\,2\frac{3}{4}$

6. $9\frac{7}{12}$
$+\,3\frac{11}{12}$

Multiply.

7. $\frac{1}{6} \times \frac{4}{5} = \underline{\quad}$

8. $\frac{3}{6} \times \frac{1}{2} = \underline{\quad}$

9. $\frac{4}{5} \times \frac{2}{7} = \underline{\quad}$

QUICK REVIEW

Three feet (3 ft or 3') equal one yard (1 yd). If you know the number of yards, multiply by three to change to feet. You will have a greater number of the smaller unit of measure.

   If you know the number of feet, divide by three to change to yards. The number will be less for the larger unit of measure. Read ft/yd as "feet per yard."

Fill in the blanks. The first two have been done for you.

10. 8 yd = __24__ ft

   8 yd × 3 ft/yd = 24 ft

11. 9 ft = __3__ yd

   9 ft ÷ 3 ft/yd = 3 yd

12. 6 ft = ____ yd

13. 5 yd = ____ ft

14. One side of a rectangle is 15 feet, and one side is 3 yards. What is the area of the rectangle in square feet? First change 3 yards to feet and then multiply to find the area.

15. What is the area of the rectangle in #14 in square yards? Change 15 feet to yards and then find the area.

16. The three sides of a triangle measure $\frac{3}{8}$ of a foot, $\frac{1}{8}$ of a foot, and $\frac{3}{4}$ of a foot. What is the perimeter of the triangle?

17. Greg is $5\frac{1}{4}$ feet tall, and Bret is $3\frac{3}{4}$ feet tall. What is the difference in their heights?

18. A cube measures three feet on a side. Give the volume in cubic yards. Change the units of measure first and then complete the problem.

SYSTEMATIC REVIEW

Add or subtract the mixed numbers. Regroup as needed.

1. $7\frac{3}{6}$
    $-\ 2\frac{5}{6}$
    _____

2. $6\frac{4}{7}$
    $-\ \ \ \frac{6}{7}$
    _____

3. 5
    $-\ 4\frac{6}{8}$
    _____

4. $2\frac{1}{2}$
    $+\ 3\frac{1}{2}$
    _____

5. $8\frac{2}{7}$
    $+\ 8\frac{6}{7}$
    _____

6. $5\frac{4}{7}$
    $+\ 2\frac{5}{7}$
    _____

Divide.

7. $\frac{7}{8} \div \frac{1}{8} =$ _____

8. $\frac{5}{9} \div \frac{2}{3} =$ _____

9. $\frac{1}{3} \div \frac{3}{4} =$ _____

10. 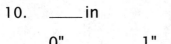 in

Fill in the blanks.

11. 24 ft = ____ yd

12. 10 yd = ____ ft

13. 36 ft = ____ yd

14. 16 yd = ____ ft

15. The base of a triangle is $\frac{1}{2}$ of a mile, and the height is $\frac{2}{5}$ of a mile. What is the area of the triangle? Multiply the two fractions and then multiply your answer by one half to find the area.

16. Rick wants to make a path 20 feet long, 4 feet wide, and $\frac{1}{2}$ of a foot thick. How many cubic feet of concrete should he order?

17. Miriam bought $3\frac{1}{8}$ pounds of red grapes and $4\frac{5}{8}$ pounds of black grapes. After the party there were $2\frac{7}{8}$ pounds of grapes left over. How many pounds of grapes were eaten?

18. If you know the area of a rectangle and the length of one dimension, you can divide to find the other dimension. If the area of a rectangle is 125 square feet and the length is 25 feet, what is the width?

19F

Add or subtract the mixed numbers. Regroup as needed.

1. $25 \frac{1}{3}$
    $- 12 \frac{1}{3}$
    _____

2. $1 \frac{5}{8}$
    $- \quad \frac{6}{8}$
    _____

3. 9
    $- 4 \frac{2}{7}$
    _____

4. $1 \frac{3}{4}$
    $+ 2 \frac{1}{4}$
    _____

5. $10 \frac{2}{6}$
    $+ 7 \frac{5}{6}$
    _____

6. $9 \frac{5}{9}$
    $+ 5 \frac{8}{9}$
    _____

Add.

7. $\frac{1}{2} + \frac{3}{7} + \frac{4}{9} = $ _____

8. $\frac{1}{2} + \frac{2}{3} + \frac{3}{5} = $ _____

9. $\frac{2}{3} + \frac{1}{4} + \frac{1}{12} = $ _____

Use the Rule of Four to make the denominators the same and then compare the fractions.

10. $\frac{3}{4}$ ◯ $\frac{4}{5}$ 11. $\frac{4}{9}$ ◯ $\frac{3}{8}$

12. $\frac{6}{11}$ ◯ $\frac{5}{12}$

Fill in the blanks.

13. 45 ft = ____ yd   14. 17 yd = ____ ft

15. 3 yd = ____ ft

16. A rectangle is $\frac{5}{8}$ of a mile long and $\frac{1}{5}$ of a mile wide. What is the area of the rectangle?

17. What is the perimeter of the rectangle in #16?

18. Lisa picked $2\frac{3}{5}$ bushels of apples. She gave $1\frac{4}{5}$ bushels to her neighbor. How much did she have left?

19. Josh ran the thirty-yard dash in gym class. How many feet did he run?

20. If the area of a rectangle is 1,260 square feet and the length is 45 feet, what is the width of the rectangle?

APPLICATION AND ENRICHMENT

A line plot is another kind of graph that may be used to compare data about different objects. The line plot below shows the results of weighing six objects. Two of the objects weighed three pounds, three of them weighed four pounds, and one weighed six pounds.

Line plots with fractions may be used to help you visualize word problems. Use the line plot below to find the information needed to answer questions 1–3.

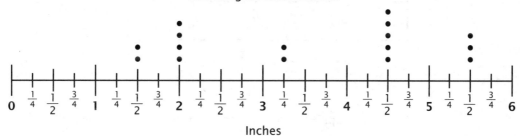

1. Isabelle gathered a collection of different kinds of leaves. She measured them and recorded each length on a line plot. How many leaves did Isabelle collect and measure?

2. What is the difference in length between the longest leaves and the shortest leaves? Find the answer by computing and by counting the spaces on the line plot. Do your answers agree?

3. Some leaves are only $1\frac{1}{2}$ inches long, and some are $4\frac{1}{2}$ inches long. If Isabelle put one of each type end to end, what would the total length of the two leaves be?

4. Use the dot plot to find how many leaves were $5\frac{1}{2}$ inches long. Change $5\frac{1}{2}$ to an improper fraction and multiply to find how long those leaves would be if placed end to end.

5. Mom had six containers with various amounts of juice in each one. The containers were all the same size. Use the information in the table to finish the line plot.

Container	Cups of Juice
#1	4
#2	$2\frac{1}{2}$
#3	$2\frac{1}{2}$
#4	4
#5	4
#6	1

Cups of Juice per Container

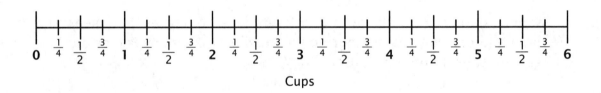

Cups

6. What amount occurred most often? What amount occurred least often?

7. What is the total amount of juice in all of the containers?

8. What is the average of the amounts of juice in all six containers?

LESSON PRACTICE

Subtract using the "same difference theorem." The first two have been done for you.

1. $6 \frac{1}{4}$ + $\frac{1}{4}$ = $6 \frac{2}{4}$

   − $2 \frac{3}{4}$ + $\frac{1}{4}$ = − 3

   $3 \frac{2}{4}$ = $3 \frac{1}{2}$

2. 6 + $\frac{2}{3}$ = $6 \frac{2}{3}$

   − $1 \frac{1}{3}$ + $\frac{2}{3}$ = − 2

   $4 \frac{2}{3}$

3. $4 \frac{1}{3}$ +   =

   − $3 \frac{2}{3}$ +   =

4. 7 +   =

   − $2 \frac{3}{4}$ +   =

5. $6 \dfrac{1}{5}$ +   =

     $- \ 3 \dfrac{2}{5}$ +   =
_____

6. 8 +   =

     $- \ 1 \dfrac{5}{8}$ +   =
_____

7. $9 \dfrac{5}{16}$ +   =

     $- \ 2 \dfrac{7}{16}$ +   =
_____

8. 10 +   =

     $- \ 4 \dfrac{7}{8}$ +   =
_____

LESSON PRACTICE

Subtract using the "same difference theorem."

1. $5 \frac{2}{5}$ +   =

    $- 1 \frac{4}{5}$ +   =

    _____

2. 9 +   =

    $- 2 \frac{3}{4}$ +   =

    _____

3. $10 \frac{1}{8}$ +   =

    $- 2 \frac{7}{8}$ +   =

    _____

4. 7 +   =

    $- 6 \frac{1}{3}$ +   =

    _____

5. $12 \frac{1}{6}$ +   =

     $- 8 \frac{2}{6}$ +   =

_____

6.   25   +   =

     $- 5 \frac{3}{5}$ +   =

_____

7. $8 \frac{3}{10}$ +   =

     $- 1 \frac{9}{10}$ +   =

_____

8.   4   +   =

     $- 3 \frac{5}{8}$ +   =

_____

LESSON PRACTICE

Subtract using the "same difference theorem."

1.　　$6\dfrac{3}{6}$ +　　=

　　$-\ 2\dfrac{5}{6}$ +　　=
　　_____

2.　　14　　+　　=

　　$-\ 5\dfrac{4}{5}$ +　　=
　　_____

3.　　$16\dfrac{2}{9}$ +　　=

　　$-\ 7\dfrac{4}{9}$ +　　=
　　_____

4.　　9　　+　　=

　　$-\ 3\dfrac{3}{4}$ +　　=
　　_____

5. $13 \frac{3}{7}$ + =

$- 8 \frac{6}{7}$ + =

6. 30 + =

$- 11 \frac{7}{10}$ + =

7. $6 \frac{1}{4}$ + =

$- 2 \frac{2}{4}$ + =

8. 5 + =

$- 1 \frac{3}{8}$ + =

SYSTEMATIC REVIEW

Subtract using the "same difference theorem."

1. $5 \frac{2}{7}$ + $\quad$ =

$- 3 \frac{6}{7}$ + $\quad$ =

_____

2. 19 $\quad$ + $\quad$ =

$- 6 \frac{2}{3}$ + $\quad$ =

_____

Add or subtract. Use the method you prefer to subtract.

3. $2 \frac{5}{8}$

$+ 1 \frac{7}{8}$

_____

4. $4 \frac{2}{5}$

$- 1 \frac{4}{5}$

_____

5. $3 \frac{1}{2}$

$+ 5 \frac{1}{2}$

_____

Find the prime factors of each number.

6. 25 _____

7. 36 _____

8. 44 _____

Multiply the numerators and denominators by the same numbers to make equivalent fractions.

9. $\dfrac{3}{7} = \dfrac{\quad}{\quad} = \dfrac{\quad}{21}$

10. $\dfrac{2}{3} = \dfrac{\quad}{\quad} = \dfrac{6}{\quad}$

11. $\dfrac{1}{9} = \dfrac{\quad}{\quad} = \dfrac{3}{\quad}$

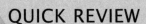

QUICK REVIEW

Two pints (pt) equal one quart (qt). Read pt/qt as "pints per quart."

Fill in the blanks. The first two have been done for you.

12. 9 qt = __18__ pt

   9 qt × 2 pt/qt = 18 pt

13. 8 pt = __4__ qt

   8 pt ÷ 2 pt/qt = 4 qt

14. 12 pt = ____ qt

15. 11 qt = ____ pt

16. Patrick bought a 20-pound bag of oranges. He has eaten $5\frac{1}{2}$ pounds of oranges. How many pounds of oranges are left?

17. Abigail made seven quarts of jam. How many pint jars does she need to hold all of the jam?

18. A cube measures six yards on a side. Give the volume in cubic feet. Change the units of measure first and then complete the problem.

SYSTEMATIC REVIEW

Subtract using the "same difference theorem."

1. $7 \frac{4}{9}$ +   =

    $- 3 \frac{7}{9}$ +   =

    _____

2. 13 +   =

    $- 7 \frac{1}{8}$ +   =

    _____

Add or subtract. Use the method you prefer to subtract.

3. $2 \frac{1}{5}$

    $+ 4 \frac{3}{5}$

    _____

4. $6 \frac{3}{6}$

    $- 4 \frac{5}{6}$

    _____

5. $5 \frac{2}{3}$

    $+ 5 \frac{2}{3}$

    _____

What is the GCF of each pair of numbers?

6. 25 and 35 _____

7. 12 and 36 _____

8. 42 and 49 _____

Subtract.

9. $\frac{1}{2} - \frac{1}{6} =$ _____

10. $\frac{7}{10} - \frac{2}{5} =$ _____

11. $\frac{11}{12} - \frac{5}{9} =$ _____

Fill in the blanks.

12. 11 qt = ____ pt   13. 18 pt = ____ qt

14. 22 pt = ____ qt

15. Seth grew $2\frac{5}{8}$ of an inch one year and $3\frac{7}{8}$ the second year. How many inches did Seth grow in those two years?

16. Is 456 divisible by 9?

17. Mitch got 16 pints of honey from his beehive. How many quarts of honey does Mitch have?

18. Use a ruler to measure the width of a book to the nearest eighth of an inch.

19. Michael earned $25 on Wednesday and $43 on Thursday. He spent $\frac{1}{4}$ of his total earnings on a gift for his sister Callie. How much did Michael spend on his sister?

20. What number is equivalent to 7^2?

SYSTEMATIC REVIEW

Subtract using the "same difference theorem."

1.
$$9 \frac{1}{6} + \quad =$$
$$- 4 \frac{5}{6} + \quad =$$

2.
$$18 \quad + \quad =$$
$$- 12 \frac{1}{2} + \quad =$$

Add or subtract. Use whichever method you prefer to subtract.

3.
$$3 \frac{3}{8}$$
$$+ 5 \frac{5}{8}$$

4.
$$8 \frac{1}{3}$$
$$- 6 \frac{1}{3}$$

5.
$$7 \frac{5}{9}$$
$$+ 7 \frac{8}{9}$$

Solve.

6. $\frac{2}{3} \times 192 = $ _____

7. $\frac{1}{5} \times 555 = $ _____

8. $\frac{3}{4} \times 84 = $ _____

Add.

9. $\frac{3}{4} + \frac{1}{2} = $ _____

10. $\frac{4}{11} + \frac{3}{10} = $ _____

11. $\frac{5}{9} + \frac{1}{3} = $ _____

Fill in the blanks.

12. 12 qt = _____ pt

13. 28 pt = _____ qt

14. 15 yd = _____ ft

15. In May we received $4\frac{2}{10}$ inches of rain. In June we received $1\frac{9}{10}$ inches less than in May. How many inches of rain did we receive in June?

16. Is 456 divisible by 3?

17. Mom bought four quarts of orange juice and six quarts of apple juice. If the juice is divided evenly among five people, how many pints of juice may each one have?

18. Measure the height of this page to the nearest eighth of an inch.

19. What is the perimeter of a square that measures $1\frac{1}{2}$ feet on each side?

20. The area of a rectangular field is $\frac{1}{8}$ square mile. One side of the field is $\frac{1}{4}$ mile long. Divide $\frac{1}{8}$ by $\frac{1}{4}$ to find the length of the other side of the field.

Numbers may be shown on a horizontal number line. On the line below, the numbers 2, 4, and 5 are marked. The arrow on the right shows us that you can make the line as long as you wish.

Number lines can also be vertical. A traditional thermometer is an example of a vertical number line. Sometimes mathematicians put together horizontal and vertical number lines to make a special kind of graph called an *x-y coordinate graph.*

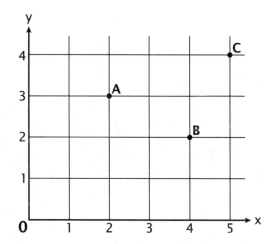

The horizontal line on the bottom of this graph is called the *x-axis*, and the vertical line on the left side is called the *y-axis*. The axes are the numbered lines on the drawing. Instead of using one number to talk about a point on a number line, two numbers are used to talk about a point on the graph. The two numbers are usually shown inside a set of parentheses and are called the *coordinates* of the point.

The point labeled A on the graph is named (2, 3). This means that you first travel two spaces to the right along the x-axis and then three spaces up along the y-axis. The point labeled B is named (4, 2). This means that you first travel four spaces to the right along the x-axis and then two spaces up along the y-axis. The point labeled C is named (5, 4) because you first travel five spaces to the right along the x-axis and then four spaces up along the y-axis.

Try your hand at graphing these points on the graph. Remember that the first coordinate is a number along the x-axis and the second coordinate corresponds to a number along the y-axis. You may want to imagine that the graph is a map of a city near you. If the coordinates are (3, 5), it means that you walk east three blocks and then turn and walk north five blocks.

Graph each set of coordinates. Mark the location with a circle and the appropriate letter.

A. (2, 4)

B. (3, 3)

C. (3, 8)

D. (5, 2)

E. (7, 3)

F. (8, 4)

G. (5, 6)

H. (7, 8)

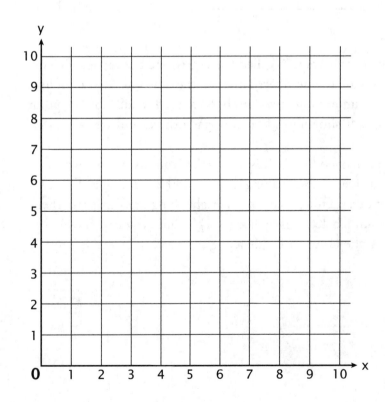

LESSON PRACTICE

Add the mixed numbers. The first one has been done for you.

1.
$$8 \frac{2}{3}$$
$$+ 5 \frac{3}{4}$$

$$\frac{8}{12} \quad \frac{2}{3} + \frac{3}{4} \quad \frac{9}{12}$$

$$8 \frac{8}{12}$$
$$+ 5 \frac{9}{12}$$
$$13 \frac{17}{12} = 14 \frac{5}{12}$$

2.
$$3 \frac{3}{8}$$
$$+ 1 \frac{4}{5}$$

3.
$$18 \frac{4}{10}$$
$$+ 3 \frac{5}{8}$$

4.
$$11 \frac{3}{5}$$
$$+ 4 \frac{5}{6}$$

5.
$$9 \frac{1}{3}$$
$$+ 6 \frac{1}{4}$$

6. $4 \dfrac{2}{3}$

   $+ \; 1 \dfrac{2}{5}$
_____

7. $9 \dfrac{3}{5}$

   $+ \; 2 \dfrac{7}{10}$
_____

8. $12 \dfrac{9}{10}$

   $+ \; 4 \dfrac{5}{8}$
_____

9. $3 \dfrac{1}{2}$

   $+ \; 1 \dfrac{3}{4}$
_____

10. $4 \dfrac{5}{8}$

   $+ \; 1 \dfrac{1}{4}$
_____

LESSON PRACTICE

Add the mixed numbers.

1. $7 \dfrac{3}{4}$
   $+ \ 9 \dfrac{3}{5}$

2. $6 \dfrac{3}{4}$
   $+ \ 6 \dfrac{5}{6}$

3. $19 \dfrac{7}{8}$
   $+ \ 2 \dfrac{5}{10}$

4. $12 \dfrac{8}{11}$
   $+ \ 5 \dfrac{2}{3}$

5. $8 \dfrac{4}{5}$
   $+ \ 2 \dfrac{1}{3}$

6. $4 \dfrac{3}{4}$
   $+ \ 2 \dfrac{1}{5}$

7. $6\dfrac{5}{9}$

    $+\ 8\dfrac{1}{3}$
_____

8. $6\dfrac{2}{3}$

    $+\ 1\dfrac{7}{9}$
_____

9. $7\dfrac{3}{10}$

    $+\ 9\dfrac{4}{5}$
_____

10. $2\dfrac{3}{5}$

    $+\ 1\dfrac{1}{2}$
_____

LESSON PRACTICE

21C

Add the mixed numbers.

1. $6\frac{2}{5}$
   $+\ 2\frac{1}{2}$

2. $5\frac{3}{7}$
   $+\ 3\frac{1}{9}$

3. $2\frac{1}{8}$
   $+\ \ \frac{4}{5}$

4. $8\frac{3}{7}$
   $+\ 7\frac{2}{9}$

5. $3\frac{4}{5}$
   $+\ 4\frac{1}{2}$

6. $2\frac{1}{3}$
   $+\ 5\frac{5}{7}$

7. $6\dfrac{3}{5}$

    $+\ \ 4\dfrac{1}{2}$
    _____

8. $1\dfrac{3}{4}$

    $+\ \ 5\dfrac{3}{5}$
    _____

9. $13\dfrac{2}{7}$

    $+\ \ 4\dfrac{1}{6}$
    _____

10. $16\dfrac{1}{2}$

    $+\ \ 17\dfrac{3}{4}$
    _____

SYSTEMATIC REVIEW

Add the mixed numbers.

1. $8\frac{5}{7}$
   $+\ 4\frac{3}{5}$
   _____

2. $6\frac{5}{7}$
   $+\ 1\frac{3}{4}$
   _____

Subtract. Use the method you prefer.

3. $5\frac{1}{6}$
   $-\ 2\frac{4}{6}$
   _____

4. $6\frac{2}{5}$
   $-\ 1\frac{1}{5}$
   _____

5. $14\frac{3}{10}$
   $-\ 9\frac{9}{10}$
   _____

Divide.

6. $\frac{1}{3} \div \frac{1}{2} =$ _____

7. $\frac{3}{6} \div \frac{2}{3} =$ _____

8. $\frac{4}{10} \div \frac{2}{5} =$ _____

Multiply.

9. $\dfrac{1}{8} \times \dfrac{6}{7} =$ _____

10. $\dfrac{4}{5} \times \dfrac{1}{9} =$ _____

11. $\dfrac{3}{5} \times \dfrac{5}{6} =$ _____

Fill in the blanks.

12. 99 ft = ____ yd

13. 6 pt = ____ qt

14. 36 qt = ____ pt

QUICK REVIEW

Use mental math to sharpen your skills. Do these problems in your head as you read them or have someone else read them aloud to you.

15. Mental Math! 5 plus 4, times 8, minus 2, divided by 10, plus 3, equals ____ .

16. Mental Math! 7 minus 3, times 6, divided by 3, times 9, equals _____ .

17. How many yards are in 13 feet? Include a fraction in the answer if you cannot divide evenly.

18. Kyle practiced the piano for $1\frac{1}{2}$ hours and the violin for $1\frac{3}{4}$ hours. For how many hours did he practice in all?

SYSTEMATIC REVIEW

Add the mixed numbers.

1.
$$8 \frac{2}{3}$$
$$+ 5 \frac{3}{4}$$

2.
$$3 \frac{3}{8}$$
$$+ 1 \frac{2}{5}$$

Subtract. Use the method you prefer.

3.
$$3 \frac{1}{4}$$
$$- 1 \frac{3}{4}$$

4.
$$7 \frac{1}{5}$$
$$- 6 \frac{4}{5}$$

5.
$$5$$
$$- 1 \frac{2}{3}$$

Divide.

6. $\frac{5}{8} \div \frac{1}{4} =$ _____

7. $\frac{3}{7} \div \frac{1}{7} =$ _____

8. $\frac{5}{9} \div \frac{2}{3} =$ _____

Multiply.

9. $\dfrac{1}{9} \times \dfrac{7}{8}$ = _____

10. $\dfrac{2}{3} \times \dfrac{1}{2}$ = _____

11. $\dfrac{7}{10} \times \dfrac{1}{5}$ = _____

Find the perimeter of each figure.

12. $3\frac{1}{4}$in $7\frac{1}{2}$ in   P = _____

13. $5\frac{1}{2}$ in $5\frac{1}{2}$ in   P = _____

14. $4\frac{1}{3}$ ft $3\frac{1}{2}$ ft $5\frac{2}{3}$ft   P = _____

15. A rectangular pool is 8 feet long, 6 feet wide, and 3 feet deep. How many cubic feet of water will it hold if it is filled to the top?

16. One twelfth of the water was drained from the pool in #15. How many cubic feet of water are left in the pool?

17. A jet was flying 17,259 feet above the Arctic Ocean. How many yards above the water was the jet?

18. How many quarts are in 25 pints? Include a fraction in the answer if you are unable to divide evenly.

19. Measure the width of your foot as accurately as you can.

20. Mental Math! 21 divided by 7, times 4, plus 3 , minus 10, equals _____ .

SYSTEMATIC REVIEW

Add the mixed numbers.

1. $18 \frac{7}{10}$
   $+ 3 \frac{5}{8}$
   —————

2. $11 \frac{3}{5}$
   $+ 4 \frac{1}{6}$
   —————

Subtract. Use the method you prefer.

3. $9 \frac{3}{7}$
   $- 5 \frac{1}{7}$
   —————

4. $17 \frac{2}{5}$
   $- 8 \frac{3}{5}$
   —————

5. 6
   $- 2 \frac{7}{8}$
   —————

Divide.

6. $\frac{6}{8} \div \frac{1}{4} =$ _____

7. $\frac{4}{5} \div \frac{1}{10} =$ _____

8. $\frac{5}{6} \div \frac{1}{12} =$ _____

Multiply.

9. $\dfrac{4}{5} \times \dfrac{1}{6} =$ _____

10. $\dfrac{5}{6} \times \dfrac{2}{8} =$ _____

11. $\dfrac{3}{5} \times \dfrac{4}{7} =$ _____

Find the area of each figure.

12. $\dfrac{1}{3}$'   A = _____

$\dfrac{2}{3}$'

13. $\dfrac{3}{4}$'   A = _____

$\dfrac{3}{4}$'

14. A = _____

2"

15. Brent did $\frac{5}{6}$ of the job, and then Alexis came along and undid $\frac{3}{5}$ of what Brent had done. What part of the whole job did Alexis undo?

16. If a cubic foot of water weighs 63 pounds, what is the weight of water that would fill a cube that measures five feet on each side?

17. How many yards are in 577 feet? Include a fraction in the answer if you cannot divide evenly.

18. Jim is 10 years old, and Ali is $5\frac{1}{2}$ years old. How much older is Jim?

19. Measure the length of your little finger as accurately as you can.

20. Mental Math! 8 times 7, minus 1, divided by 5, minus 3, equals _____ .

Think of this graph as a map of the streets in Bill's town. Use the graph to help you answer the questions about Bill's travels.

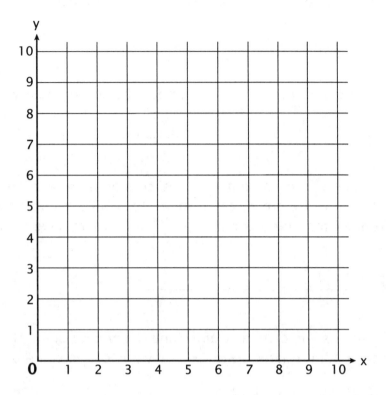

1. Bill's house is at (0, 0). Beginning at that point, Bill walked two blocks east (to the right) and three blocks north (towards the top) to the grocery store. Make a dot at the point where he ended his walk and label it G. What are the coordinates of G?

2. When Bill left the grocery store, he walked three blocks east and then four blocks north to the cafe. Make a dot to show the cafe and label it C. What are the coordinates of C? How many blocks has Bill walked so far?

3. Draw lines on the graph that show Bill's route from the beginning of his walk to the cafe. Choose a different route home to (0, 0) for him and trace it on the graph. How many blocks must he walk to follow the new route home?

Graphs may also be used to show how different number patterns are related to each other. Follow the directions for an example of this.

4. Start with zero and fill in the blanks by adding 2 each time.

    0 , _____ , _____ , _____ , _____

5. Start with zero and make a list of numbers by adding 4 each time.

    _____ , _____ , _____ , _____ , _____

6. Use the numbers from the patterns you created in #4 and #5 to make coordinate pairs. Use the values from #4 for your x-coordinates and the values from #5 for your y-coordinates. The first two have been done for you.

    (0, 0) , (2, 4), _____ , _____ , _____

7. Graph the points and draw a straight line to connect them. How is the y-axis value of each point related to the x-axis value of the same point?

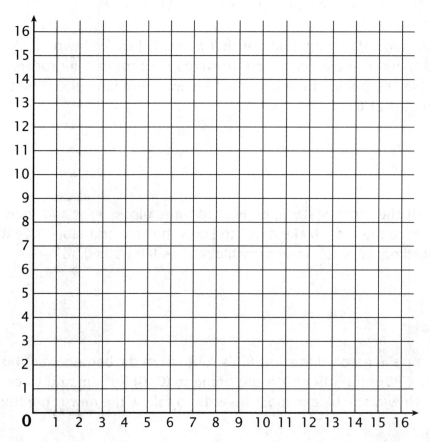

Subtract the mixed numbers and regroup as necessary. Use the traditional method or the "same difference theorem" to subtract once the denominators are the same. The first two have been done for you.

1. $6 \frac{4}{5}$ $\frac{12}{15}$ $\frac{4}{5} - \frac{1}{3}$ $\frac{5}{15}$ $6 \frac{12}{15}$

   $- 5 \frac{1}{3}$ $- 5 \frac{5}{15}$

                                                  $1 \frac{7}{15}$

2. $7 \frac{1}{4}$ $\frac{2}{8}$ $\frac{1}{4} - \frac{1}{2}$ $\frac{4}{8}$ $^6\cancel{7} \frac{10}{8}$

   $- 6 \frac{1}{2}$ $- 6 \frac{4}{8}$

                                                  $\frac{6}{8} = \frac{3}{4}$

3. $9 \frac{1}{3}$

   $- 6 \frac{3}{8}$

4. $10 \frac{5}{6}$

   $- 3 \frac{1}{3}$

5. $7 \dfrac{3}{5}$
   $- \ 2 \dfrac{8}{9}$
   _____

6. $9 \dfrac{1}{8}$
   $- \ 8 \dfrac{1}{2}$
   _____

7. $8 \dfrac{1}{2}$
   $- \ 2 \dfrac{3}{4}$
   _____

8. $5 \dfrac{1}{3}$
   $- \ 2 \dfrac{5}{6}$
   _____

9. $11 \dfrac{1}{3}$
   $- \ 3 \dfrac{1}{4}$
   _____

10. $26 \dfrac{2}{3}$
    $- \ 22 \dfrac{4}{5}$
    _____

Subtract the mixed numbers and regroup as necessary. Use the method you prefer to subtract.

1. $10 \dfrac{1}{4}$
    $- 3 \dfrac{4}{7}$
    _____

2. $9 \dfrac{1}{3}$
    $- 2 \dfrac{5}{6}$
    _____

3. $7 \dfrac{4}{5}$
    $- 7 \dfrac{1}{2}$
    _____

4. $3 \dfrac{1}{3}$
    $- 1 \dfrac{4}{5}$
    _____

5. $5 \dfrac{1}{6}$
    $- 2 \dfrac{4}{5}$
    _____

6. $8 \dfrac{3}{5}$
    $- 4 \dfrac{1}{2}$
    _____

7.　　$15 \dfrac{5}{7}$

　　$-\,12 \dfrac{3}{4}$
　　————

8.　　$2 \dfrac{1}{2}$

　　$-\,2 \dfrac{1}{4}$
　　————

9.　　$7 \dfrac{3}{10}$

　　$-\,5 \dfrac{2}{3}$
　　————

10.　　$8 \dfrac{1}{2}$

　　$-\,4 \dfrac{2}{9}$
　　————

Subtract the mixed numbers and regroup as necessary. Use the method you prefer to subtract.

1. $8\dfrac{1}{6}$
   $-1\dfrac{3}{10}$

2. $5\dfrac{1}{4}$
   $-2\dfrac{5}{6}$

3. $14\dfrac{3}{4}$
   $-11\dfrac{7}{8}$

4. $7\dfrac{2}{5}$
   $-3\dfrac{1}{4}$

5. $8\dfrac{1}{6}$
   $-2\dfrac{2}{3}$

6. $10\dfrac{1}{2}$
   $-3\dfrac{4}{5}$

7. $8 \dfrac{5}{7}$

$- \; 2 \dfrac{1}{2}$

8. $4 \dfrac{1}{2}$

$- \; 1 \dfrac{3}{4}$

9. $6 \dfrac{4}{7}$

$- \; 4 \dfrac{2}{3}$

10. $10 \dfrac{1}{6}$

$- \; 5 \dfrac{2}{5}$

SYSTEMATIC REVIEW

Add or subtract the mixed numbers.

1. $8 \frac{2}{7}$
   $- 2 \frac{3}{4}$

2. $14 \frac{1}{3}$
   $- 3 \frac{1}{2}$

3. $6 \frac{4}{5}$
   $+ 1 \frac{2}{3}$

4. $4 \frac{3}{5}$
   $+ 1 \frac{2}{5}$

5. $12 \frac{1}{8}$
   $+ 9 \frac{4}{7}$

Multiply or divide.

6. $\frac{1}{2} \div \frac{1}{2} = $ _____

7. $\frac{1}{4} \times \frac{5}{6} = $ _____

8. $\frac{5}{6} \div \frac{7}{8} = $ _____

Convert each mixed number to an improper fraction.

9. $6\frac{2}{3}$ = _____

10. $27\frac{4}{9}$ = _____

11. $13\frac{3}{5}$ = _____

QUICK REVIEW

Four quarts (qt) equal one gallon (gal).

Fill in the blanks. Include a fraction in the answer if you are unable to divide evenly. The first two have been done for you. Read qt/gal as "quarts per gallon."

12. 17 qt = $4\frac{1}{4}$ gal

   17 qt ÷ 4 qt/gal = $4\frac{1}{4}$ gal

13. 8 gal = __32__ qt

   8 gal × 4 qt/gal = 32 qt

14. 10 gal = ____ qt

15. 11 qt = ____ gal

16. A turtle crawled $4\frac{3}{10}$ yards and then turned around and crawled back $3\frac{9}{10}$ yards. How far from its starting point did the turtle finish?

17. What are the dimensions (length and width) of a dollar bill to the nearest eighth of an inch?

18. Mental Math! 9 minus 1, times 7, plus 4, divided by 6, equals _____ .

SYSTEMATIC REVIEW

Add or subtract the mixed numbers.

1. $10 \dfrac{1}{8}$
    $- 4 \dfrac{6}{7}$

2. $4 \dfrac{1}{5}$
    $- 3 \dfrac{3}{4}$

3. $2 \dfrac{1}{10}$
    $+ 6 \dfrac{1}{5}$

4. $2 \dfrac{7}{8}$
    $+ 1 \dfrac{2}{3}$

5. $25 \dfrac{2}{9}$
    $+ 7 \dfrac{5}{8}$

Multiply or divide.

6. $\dfrac{1}{5} \times \dfrac{1}{10} =$ _____

7. $\dfrac{1}{5} \div \dfrac{1}{10} =$ _____

8. $\dfrac{1}{2} \times \dfrac{3}{5} =$ _____

EPSILON SYSTEMATIC REVIEW 22E

307

Use the Rule of Four to make denominators the same. Then compare the fractions.

9. $\dfrac{6}{8} \bigcirc \dfrac{3}{4}$ 10. $\dfrac{1}{10} \bigcirc \dfrac{2}{9}$

11. $\dfrac{5}{11} \bigcirc \dfrac{3}{8}$

Fill in the blanks. Include a fraction in the answer if the numbers do not divide evenly.

12. 23 qt = ____ gal   13. 5 gal = ____ qt

14. 30 yd = ____ ft

15. Five and one half gallons of lemonade and three and one half gallons of punch were brought to the picnic. The guests drank six gallons of punch or lemonade. How many *quarts* of drink were left over to take home?

16. Nicholas had $\frac{5}{16}$ of a cake left from his birthday. He divided the cake into pieces that were each $\frac{1}{8}$ of a cake. How many pieces does he have now? Notice that your answer does not come out evenly. One piece will be half the size of the others.

17. What are the prime factors of 48?

18. What number is equivalent to 4^2?

19. Measure the distance across a quarter to the nearest eighth of an inch.

20. Mental Math! 7 plus 2, times 7, plus 2, minus 61, equals _____ .

SYSTEMATIC REVIEW

Add or subtract the mixed numbers.

1. $12 \frac{7}{9}$
    $- 1 \frac{1}{2}$

2. $5 \frac{1}{3}$
    $- 2 \frac{2}{3}$

3. $1 \frac{1}{2}$
    $+ 3 \frac{2}{3}$

4. $4 \frac{2}{3}$
    $+ 3 \frac{4}{5}$

5. $6 \frac{3}{4}$
    $+ 9 \frac{7}{8}$

Multiply or divide.

6. $\frac{4}{5} \div \frac{1}{3} = $ _____

7. $\frac{3}{5} \times \frac{3}{7} = $ _____

8. $\frac{1}{4} \div \frac{3}{4} = $ _____

EPSILON SYSTEMATIC REVIEW 22F

Convert each mixed number to an improper fraction.

9. $7\frac{3}{4}$ = _____

10. $10\frac{1}{8}$ = _____

11. $45\frac{1}{2}$ = _____

Fill in the blanks. Include a fraction in the answer if you are unable to divide evenly.

12. 24 qt = ____ gal

13. 20 gal = ____ qt

14. 15 pt = ____ qt

15. The area of a rectangle is 315 square feet. It is 35 feet long. What is the width of the rectangle?

16. Give the width of the rectangle in #15 in yards.

17. Is 356 divisible by 9?

18. What is the GCF of 15 and 45?

19. There are a total of 24 people in the room. Three fourths of the people in the room want to go home. Two thirds of those who want to go home have a ride. How many of those who want to go home *do not* have a ride and will have to walk? (Read this carefully and work it one step at a time.)

20. Mental Math! 8 times 6, minus 8, divided by 4, times 3, equals _____ .

Parentheses tell you what part of a problem should be calculated first.

$6 \times (2 + 4) =$

$6 \times (2 + 4) = 6 \times (6) = 36$

More complex problems may use brackets. Compute the numbers inside the parentheses first and then compute the numbers inside the brackets. Remember that brackets are just another type of parentheses.

$$[2 \times (2 + 1)] - 4 =$$

$$[2 \times \ (3) \ ] - 4 =$$

$$[\ 6 \ ] - 4 = 2$$

You may encounter another type of grouping symbol known as braces in more elaborate expressions and equations. As always, compute the numbers inside the parentheses first; then compute the numbers inside the brackets, followed by those inside the braces. (Braces are sometimes called "curly brackets.")

$$3 \times \{[(6 - 4) \times 5] - 1\} =$$

$$3 \times \{[(2) \times 5] - 1\} =$$

$$3 \times \{[10] - 1\} =$$

$$3 \times \{9\} = 27$$

Pay close attention to the grouping symbols. Compute each equation one step at a time.

1. $\{[(36 \div 6) + 3] - 1\} \times 2 =$

2. $\{[(17 - 9) \times 3] - 2\} \div 11 =$

3. $2 \times \{[8 + (4 \times 7)] \div 4\} =$

4. $5 + \{[3(6 + 1)] - 4\} =$

An important part of learning about math is being able to see patterns in numbers. Study each chart below and see if you can determine the pattern. Once you have found the pattern, use it to fill in the empty spaces in the chart. The first one has instructions to guide you.

5.

2	4		8		12		16
3	6	9		15		21	

Count by two to get the numbers in the top line.

Count by three to get the numbers in the bottom line.

Add a number that is one greater each time to the top numbers to get the numbers in the bottom row.

6.

1			4		6		8
4	8			20			

7.

1	2		7	11		22	29
0	1	3		10			28

8.

3		9					24
	21	18		12		6	3

Multiply each fraction by its reciprocal. The first one has been done for you.

1. $\dfrac{5}{6} \times \dfrac{6}{5} = \dfrac{30}{30} = 1$

2. $\dfrac{3}{4} \times \dfrac{4}{3} = \underline{\hspace{1.5cm}} = \underline{\hspace{1.5cm}}$

3. $\dfrac{1}{2} \times \dfrac{2}{1} = \underline{\hspace{1.5cm}} = \underline{\hspace{1.5cm}}$

Write the reciprocal of each number. The first two have been done for you.

4. $\dfrac{2}{3}$ $\quad \dfrac{3}{2}$ $\underline{\hspace{2cm}}$

5. 4 $\quad \dfrac{1}{4}$ $\underline{\hspace{2cm}}$

6. $\dfrac{7}{8}$ $\underline{\hspace{2cm}}$

Divide using the reciprocal. Then check your work by using the Rule of Four. The first two have been done for you.

reciprocal

7. $\dfrac{1}{2} \div \dfrac{5}{9} = \dfrac{1}{2} \times \dfrac{9}{5} = \dfrac{9}{10}$

Rule of Four

8. $\dfrac{1}{2} \div \dfrac{5}{9} = \dfrac{9}{18} \div \dfrac{10}{18} = \dfrac{9}{10}$

9. $\dfrac{3}{4} \div \dfrac{5}{8} =$

10. $\dfrac{3}{4} \div \dfrac{5}{8} =$

11. $\dfrac{9}{10} \div \dfrac{1}{5} =$

12. $\dfrac{9}{10} \div \dfrac{1}{5} =$

Convert the mixed numbers to improper fractions. Then divide using the reciprocal. The first two have been done for you.

13. $1\dfrac{1}{2} \div 5\dfrac{1}{3} = \dfrac{3}{2} \times \dfrac{3}{16} = \dfrac{9}{32}$

14. $3\dfrac{1}{3} \div \dfrac{5}{18} = \dfrac{10}{3} \times \dfrac{18}{5} = \dfrac{180}{15} = 12$

15. $1\dfrac{5}{6} \div 3\dfrac{1}{4} =$

16. $1\dfrac{7}{8} \div \dfrac{5}{8} =$

17. A rectangle has an area of $1\frac{2}{3}$ square miles. The rectangle is $\frac{5}{9}$ of a mile wide. How long is the rectangle? ($1\frac{2}{3} \div \frac{5}{9}$)

18. There were $2\frac{2}{3}$ pies left after the family reunion. How many people can be served if each piece is $\frac{1}{6}$ of a pie? ($2\frac{2}{3} \div \frac{1}{6}$)

Multiply each fraction by its reciprocal.

1. $\dfrac{1}{3} \times \dfrac{3}{1} = $ —— $=$ ____

2. $\dfrac{5}{8} \times \dfrac{8}{5} = $ —— $=$ ____

3. $\dfrac{3}{7} \times \dfrac{7}{3} = $ —— $=$ ____

Write the reciprocal of each number.

4. $\dfrac{1}{2}$ _____

5. 9 _____

6. $\dfrac{3}{4}$ _____

Divide using the reciprocal. Then check your work by using the Rule of Four.

reciprocal

Rule of Four

7. $\dfrac{4}{1} \div \dfrac{1}{2} = $

8. $\dfrac{4}{1} \div \dfrac{1}{2} = $

9. $\dfrac{7}{10} \div \dfrac{7}{12} = $

10. $\dfrac{7}{10} \div \dfrac{7}{12} = $

11. $\dfrac{3}{4} \div \dfrac{1}{8} = $

12. $\dfrac{3}{4} \div \dfrac{1}{8} = $

Change the mixed numbers to improper fractions. Then divide using the reciprocal.

13. $7\frac{3}{5} \div 1\frac{1}{2} =$

14. $3\frac{1}{4} \div 10\frac{4}{7} =$

15. $1\frac{1}{2} \div \frac{1}{8} =$

16. $2\frac{1}{6} \div \frac{1}{6} =$

17. A rectangle has an area of $1\frac{1}{5}$ square feet. One dimension of the rectangle is $\frac{9}{10}$ of a foot. What is the other dimension of the rectangle?

18. Mom has five pounds of ground beef in the freezer. How many times can she make a recipe that requires $1\frac{1}{4}$ pounds of ground beef? $(\frac{5}{1} \div 1\frac{1}{4})$

Multiply each fraction by its reciprocal.

1. $\dfrac{1}{6} \times \dfrac{6}{1} = \underline{\quad\quad} = \underline{\quad\quad}$

2. $\dfrac{2}{3} \times \dfrac{3}{2} = \underline{\quad\quad} = \underline{\quad\quad}$

3. $\dfrac{4}{5} \times \dfrac{5}{4} = \underline{\quad\quad} = \underline{\quad\quad}$

Write the reciprocal of each number.

4. $\dfrac{5}{6}$ _____

5. 13 _____

6. $\dfrac{9}{4}$ _____

Divide using the reciprocal and then check your work by using the Rule of Four.

reciprocal

Rule of Four

7. $\dfrac{5}{7} \div \dfrac{2}{9} =$

8. $\dfrac{5}{7} \div \dfrac{2}{9} =$

9. $\dfrac{7}{8} \div \dfrac{1}{3} =$

10. $\dfrac{7}{8} \div \dfrac{1}{3} =$

11. $\dfrac{2}{1} \div \dfrac{2}{3} =$

12. $\dfrac{2}{1} \div \dfrac{2}{3} =$

Change the mixed numbers to improper fractions. Then divide using the reciprocal.

13. $2\frac{2}{3} \div 1\frac{3}{8} =$

14. $5\frac{1}{3} \div 2\frac{2}{3} =$

15. $2\frac{4}{5} \div \frac{1}{10} =$

16. $1\frac{3}{4} \div \frac{5}{12} =$

17. George ran for $2\frac{1}{2}$ miles. He divided his run into one-half-mile sections. How many sections were there?

18. Sally has $6\frac{1}{4}$ yards of fabric. She wants to make gifts for her five grandchildren and plans to use the same amount of fabric for each gift. How many yards will she use for each one if she uses the whole $6\frac{1}{4}$ yards?

Divide using the reciprocal and then check your work by using the Rule of Four.

reciprocal

Rule of Four

1. $\dfrac{1}{3} \div \dfrac{1}{5} =$

2. $\dfrac{1}{3} \div \dfrac{1}{5} =$

3. $\dfrac{3}{4} \div \dfrac{5}{8} =$

4. $\dfrac{3}{4} \div \dfrac{5}{8} =$

Change the mixed numbers to improper fractions. Then divide using the reciprocal.

5. $2\dfrac{1}{4} \div \dfrac{3}{5} =$

6. $1\dfrac{5}{6} \div 2\dfrac{1}{10} =$

Add or subtract the mixed numbers.

7. $\begin{array}{r} 9\dfrac{1}{6} \\ -\ 4\dfrac{2}{5} \\ \hline \end{array}$

8. $\begin{array}{r} 25\dfrac{1}{3} \\ -\ 16\dfrac{2}{3} \\ \hline \end{array}$

9. $\begin{array}{r} 7\dfrac{5}{8} \\ +\ 2\dfrac{1}{4} \\ \hline \end{array}$

QUICK REVIEW

Sixteen ounces (oz) equals one pound (lb).

Fill in the blanks. Include a fraction in the answer if you cannot divide evenly. The first two have been done for you. Read oz/lb as "ounces per pound."

10. 32 oz = __2__ lb

    32 oz ÷ 16 oz/lb = 2 lb

11. 3 lb = __48__ oz

    3 oz × 16 oz/lb = 48 lb

12. 25 oz = ____ lb

13. 10 lb = ____ oz

14. Jacob has a rope that is $5\frac{5}{6}$ feet long. How many $1\frac{1}{6}$-foot long pieces may be cut from the rope?

15. Peggy bought $5\frac{1}{2}$ pounds of raisins and $4\frac{3}{4}$ pounds of dates. After using $6\frac{3}{8}$ pounds of fruit to make fruitcakes, how many pounds of fruit did Peggy have left over?

16. Matthew dug a hole eight feet long, four feet wide, and two feet deep. How many cubic feet of dirt did he take out of the hole?

17. The three sides of a triangle measure $\frac{1}{2}$ ft, $\frac{1}{2}$ ft, and $\frac{3}{4}$ ft. What is the perimeter of the triangle?

18. Mental Math! 6 plus 3, minus 4, plus 2, times 7, equals _____ .

Divide using the reciprocal and then check your work by using the Rule of Four.

reciprocal   Rule of Four

1. $\dfrac{2}{3} \div \dfrac{5}{7} =$ 2. $\dfrac{2}{3} \div \dfrac{5}{7} =$

3. $\dfrac{9}{10} \div \dfrac{1}{5} =$ 4. $\dfrac{9}{10} \div \dfrac{1}{5} =$

Change the mixed numbers to improper fractions. Then divide using the reciprocal.

5. $4\dfrac{3}{5} \div 2\dfrac{3}{7} =$ 6. $1\dfrac{4}{5} \div 1\dfrac{1}{3} =$

Add or subtract the mixed numbers.

7. $\begin{array}{r} 4\frac{3}{4} \\ -\ 1\frac{1}{2} \\ \hline \end{array}$ 8. $\begin{array}{r} 13\frac{2}{7} \\ -\ 11\frac{1}{3} \\ \hline \end{array}$

9. $\begin{array}{r} 3\frac{1}{3} \\ +\ 2\frac{2}{3} \\ \hline \end{array}$

Add.

10. $\dfrac{3}{4} + \dfrac{3}{5} + \dfrac{6}{10} =$ _____

11. $\dfrac{1}{3} + \dfrac{5}{7} + \dfrac{2}{5} =$ _____

12. $\dfrac{1}{2} + \dfrac{2}{3} + \dfrac{4}{7} =$ _____

Fill in the blanks. Include a fraction in the answer if you cannot divide evenly.

13. 64 oz = _____ lb

14. 6 lb = _____ oz

15. 13 qt = _____ gal

16. What is the area of a triangle with a base of 20 feet and a height of 5 feet?

17. Arnold painted $10\frac{1}{2}$ feet of fence in the morning and $7\frac{1}{2}$ feet in the afternoon. How many yards of fence has he painted? (Add first and then change to yards.)

18. The area of a rectangle is $9\frac{3}{10}$ square feet. One dimension is $2\frac{2}{5}$ feet. What is the other dimension of the rectangle?

19. What is the volume of a cube if each side is seven feet long?

20. Mental Math! 11 minus 4, plus 3, plus 12, times 2, equals _____ .

SYSTEMATIC REVIEW

Divide using the reciprocal and then check your work by using the Rule of Four.

reciprocal   Rule of Four

1. $\dfrac{5}{9} \div \dfrac{2}{3} =$ 2. $\dfrac{5}{9} \div \dfrac{2}{3} =$

3. $\dfrac{1}{2} \div \dfrac{1}{3} =$ 4. $\dfrac{1}{2} \div \dfrac{1}{3} =$

Change the mixed numbers to improper fractions. Then divide using the reciprocal.

5. $2\dfrac{3}{4} \div \dfrac{5}{8} =$ 6. $2\dfrac{2}{5} \div 1\dfrac{1}{6} =$

Add or subtract the mixed numbers.

7. $\begin{array}{r} 12\dfrac{1}{7} \\ -\ 6\dfrac{4}{7} \\ \hline \end{array}$ 8. $\begin{array}{r} 3\dfrac{1}{8} \\ -\ 2\dfrac{1}{2} \\ \hline \end{array}$

9. $\begin{array}{r} 5\dfrac{1}{10} \\ +\ 2\dfrac{2}{5} \\ \hline \end{array}$

Multiply the numerators and denominators by the same number to make equivalent fractions.

10. $\dfrac{3}{4} = \underline{\quad} = \dfrac{\quad}{12}$

11. $\dfrac{1}{8} = \underline{\quad} = \dfrac{3}{\quad}$

12. $\dfrac{7}{10} = \underline{\quad} = \dfrac{21}{\quad}$

Fill in the blanks. Include a fraction in the answer if you cannot divide evenly.

13. 21 oz = _____ lb

14. 14 pt = _____ qt

15. 25 yd = _____ ft

16. Jane has $3\frac{3}{4}$ pounds of walnuts. If she divides them evenly among five people, how much will each person get? (Write 5 as $\frac{5}{1}$.)

17. A room is 14 feet by 11 feet. The ceiling is nine feet high. How many cubic feet of air does the room hold?

18. What is the reciprocal of 10?

19. Ryan put 20 gallons of gasoline in his empty tank. He has used $\frac{3}{4}$ of what he put in the tank. How many quarts of gasoline are left in his tank?

20. Mental Math! 15 divided by 5, plus 5, times 6, plus 2, equals _____ .

APPLICATION AND ENRICHMENT

In this lesson you learned how to solve a division problem with fractions by multiplying by the reciprocal. Studying word problems and thinking about their meaning can help you understand why this strategy works.

Example 1

Tom had eight jelly beans. He divided them into two parts to share with a friend. How many jelly beans did each person get?

$8 \div 2 = 4$ jelly beans

$8 \times \frac{1}{2} = 4$ jelly beans

(You may also think of this as $\frac{1}{2}$ of 8.)

Example 2

A path is $\frac{4}{5}$ of a mile long. If it is divided in sections that are each $\frac{1}{5}$ of a mile long, how many sections will there be?

$\frac{4}{5} \div \frac{1}{5} = 4 \div 1 = 4$ sections

$\frac{4}{5} \times \frac{5}{1} = \frac{20}{5} = 20 \div 5 = 4$ sections

(You may also think of this as $\frac{4}{5}$ of 5.)

Example 3

A rectangular shape has an area of $2\frac{1}{4}$ square feet. The length of one side is $1\frac{1}{2}$ feet. What is the length of the other side of the shape? Notice how the parts of the square shown in the drawing add up to the total area.

$2\frac{1}{4} \div 1\frac{1}{2} = 9/4 \div 3/2$

Using the Rule of Four:

$\frac{18}{8} \div \frac{12}{8} = 18 \div 12 = \frac{18}{12} = 1\frac{6}{12} = 1\frac{1}{2}$ ft

Multiplying by the inverse:

$\frac{9}{4} \times \frac{2}{3} = \frac{18}{12} = 1\frac{6}{12} = 1\frac{1}{2}$ ft

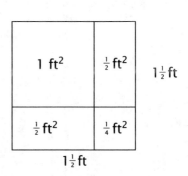

Study each of the division word problems. Solve each one using the method you prefer. Draw each problem to show that the answer makes sense.

1. Mary has a piece of string that is $\frac{5}{6}$ of a yard long. How many $\frac{1}{6}$-yard long pieces can she cut from it?

2. One third of a pizza is left in the pan. If the pizza is divided among four people, what part of a whole pizza will each person have?

3. A rectangular shape has an area of $3\frac{1}{8}$ square feet. The length of one side is $1\frac{1}{4}$ feet. What is the length of the other side of the shape? Find the missing side first and then draw the rectangle to see if the drawing shows the given area.

4. Mom has six cups of raisins. How many times can she make a recipe that calls for $\frac{1}{2}$ of a cup of raisins?

Write the multiplicative inverse of each number. The first two have been done for you.

1. $\dfrac{1}{4}$ $\dfrac{4}{1}$ _____

2. 5 $\dfrac{1}{5}$ _____

3. $\dfrac{9}{10}$ _____

4. 7 _____

Use the multiplicative inverse to solve each equation. Check your work by replacing the unknown with the solution. The first two have been done for you.

5. $6X = 36$

$$6X = 36$$
$$\frac{1}{6} \cdot \frac{6X}{1} = \frac{36}{1} \cdot \frac{1}{6}$$
$$X = 6$$

6. Check for #5
$$6X = 36$$
$$6(6) = 36$$
$$36 = 36$$

7. $8A = 40$

8. Check for #7

9. $3D = 27$

10. Check for #9

11. $11Y = 121$

12. Check for #11

13. Julie has R rabbits. If she had three times that number of rabbits, she would have 12. How many rabbits does Julie have now? (3R = 12)

14. Sam has D dollars. Sarah has four times as much money as Sam. If Sarah has 48 dollars, how much money does Sam have? (4D = 48)

LESSON PRACTICE

Write the multiplicative inverse of each number.

1. $\frac{2}{3}$ _____

2. 10 _____

3. $\frac{1}{8}$ _____

4. 2 _____

Use the multiplicative inverse to solve each equation. Check your work by replacing the unknown with the solution.

5. $8Z = 72$

6. Check for #5

7. $10C = 100$

8. Check for #7

9. $5F = 25$

10. Check for #9

11. $2A = 6$

12. Check for #11

13. Two times Megan's age is 12. How old is Megan? (2A = 12)

14. Ethan read B books. Isaac read five times as many books as Ethan did. If Isaac read 15 books, how many did Ethan read? (5B = 15)

LESSON PRACTICE

LESSON PRACTICE

Write the multiplicative inverse of each number.

1. $\dfrac{7}{9}$ _____

2. 8 _____

3. $\dfrac{3}{2}$ _____

4. 32 _____

Use the multiplicative inverse to solve each equation. Check your work by replacing the unknown with the solution.

5. $4B = 8$

6. Check for #5

7. $5E = 20$

8. Check for #7

9. $2H = 24$

10. Check for #9

11. $3C = 63$

12. Check for #11

13. Five times Andrew's age is 50. How old is Andrew? Write the equation and solve. (If no letter is given, you may choose any one you like.)

14. Daniel hopped X feet. Aaron hopped nine times as far as Daniel. If Aaron hopped 81 feet, how far did Daniel hop? Write the equation and solve.

SYSTEMATIC REVIEW

Use the multiplicative inverse to solve each equation. Check your work by replacing the unknown with the solution.

1. $7F = 49$

2. Check for #1

3. $12R = 36$

4. Check for #3

Divide using the reciprocal and then check your work by using the Rule of Four.

reciprocal

Rule of Four

5. $\dfrac{1}{2} \div \dfrac{1}{4} =$

6. $\dfrac{1}{2} \div \dfrac{1}{4} =$

Change the mixed numbers to improper fractions. Then divide using the reciprocal.

7. $\dfrac{2}{5} \div 1\dfrac{5}{6} =$

8. $5\dfrac{1}{2} \div 2\dfrac{4}{9} =$

Add or subtract the mixed numbers.

9. $\begin{array}{r} 8\frac{2}{7} \\ -\ 3\frac{4}{5} \\ \hline \end{array}$

10. $\begin{array}{r} 16\frac{5}{8} \\ -\ 10\frac{3}{8} \\ \hline \end{array}$

11.
$$5\,\frac{7}{10}$$
$$+\ 4\,\frac{1}{2}$$
_____

QUICK REVIEW

Twelve inches (in) equal one foot (ft).

Fill in the blanks. Include a fraction in the answer if you cannot divide evenly. The first two have been done for you. Read in/ft as "inches per foot."

12. 25 in = $2\frac{1}{2}$ ft

    25 in ÷ 12 in/ft =

    $\dfrac{25\,\text{in}}{1} \times \dfrac{1\,\text{ft}}{12\,\text{in}} = 2\frac{1}{12}$ in

13. 3 ft = __36__ in

    3 ft × 12 in/ft = 36 in

14. 60 in = ____ ft

15. 8 ft = ____ in

16. Jim wrapped X Christmas gifts for his children. Lisa wrapped four times as many gifts as Jim. If Lisa wrapped 16 gifts, how many gifts did Jim wrap? Write the equation and solve.

17. Linda received a five-pound box of chocolates from her husband. How many ounces of chocolate does she have?

18. Mental Math! 9 times 9, minus 1, divided by 2, divided by 5, equals _____ .

Use the multiplicative inverse to solve each equation. Check your work by replacing the unknown with the solution.

1. $6G = 42$

2. Check for #1

3. $9S = 54$

4. Check for #3

Divide using the reciprocal and then check your work by using the Rule of Four.

reciprocal

5. $\dfrac{6}{8} \div \dfrac{1}{2} =$

Rule of Four

6. $\dfrac{6}{8} \div \dfrac{1}{2} =$

Change the mixed numbers to improper fractions. Then divide using the reciprocal.

7. $6\dfrac{4}{5} \div 1\dfrac{2}{3} =$

8. $8\dfrac{1}{3} \div 5\dfrac{5}{8} =$

Add or subtract the mixed numbers.

9. $4\dfrac{1}{4}$
$+\ 2\dfrac{3}{4}$

10. $2\dfrac{1}{10}$
$+\ 6\dfrac{1}{5}$

11. $13\frac{1}{2}$
    $-\ 5\frac{7}{9}$
    _____

Multiply.

12. $\frac{5}{6} \times \frac{2}{3} =$ _____

13. $\frac{4}{5} \times \frac{2}{7} =$ _____

14. $\frac{1}{9} \times \frac{3}{8} =$ _____

15. Three times Sue's age is 45. How old is Sue? Write the equation and solve.

16. A square measures $\frac{5}{8}$ of an inch on each side. Give the area and the perimeter of the square.

17. Michael measured 32 feet for a path to his front door. How many yards long is his path?

18. How many inches long is Michael's path? (See #17.)

19. What number is equivelent to 15^2?

20. Mental Math! 30 minus 10, minus 2, divided by 6, times 7, equals _____ .

SYSTEMATIC REVIEW

24F

Use the multiplicative inverse to solve each equation. Check your work by replacing the unknown with the solution.

1. $8H = 72$

2. Check for #1

3. $11T = 55$

4. Check for #3

Divide using the reciprocal and then check your work by using the Rule of Four.

reciprocal

5. $\dfrac{1}{4} \div \dfrac{1}{3} =$

Rule of Four

6. $\dfrac{1}{4} \div \dfrac{1}{3} =$

Change mixed numbers to improper fractions. Then divide using the reciprocal.

7. $3\dfrac{1}{2} \div 4\dfrac{1}{3} =$

8. $4\dfrac{3}{5} \div 1\dfrac{2}{3} =$

Add or subtract the mixed numbers.

9. $8\dfrac{3}{4}$

   $-\ 8\dfrac{1}{4}$

10. $3\dfrac{7}{8}$

   $+\ 2\dfrac{1}{2}$

11.
$$7\frac{2}{5}$$
$$-\ 3\frac{5}{8}$$
_____

Multiply.

12. $\dfrac{3}{4} \times \dfrac{1}{8} =$ _____

13. $\dfrac{1}{3} \times \dfrac{3}{7} =$ _____

14. $\dfrac{2}{5} \times \dfrac{5}{6} =$ _____

15. Christina has D dimes. Amanda has six times as many dimes as Christina. If Amanda has 18 dimes, how many does Christina have? Write the equation and solve.

16. A rectangle measures 5 feet by 10 feet. What is the perimeter in yards?

17. Debbie made $5\frac{1}{2}$ gallons of punch for her party. Her guests drank $4\frac{3}{4}$ gallons. How many gallons are left over?

18. There are 12 inches in one foot. How many inches are in $\frac{3}{4}$ of a foot?

19. How many quarts are in $\frac{3}{4}$ of a gallon?

20. Mental Math! 20 divided by 4, plus 4, times 2, divided by 6, equals _____ .

You have been reviewing measurements regularly in your student book. Here is another way to look at converting one measure to another.

David produced the following amounts of maple syrup last week. He recorded the amounts in a table as gallons, but most customers want to buy only a quart of syrup.

	gallons	quarts
Monday	10	
Tuesday	15	
Wednesday	8	
Thursday	4	
Friday	12	

1. How many quarts are in one gallon?

2. Should you multiply or divide by your answer to number 1 in order to change gallons to quarts?

3. Use your answers to numbers 1 and 2 to fill in the table.

4. On which day was the most maple syrup produced? What was the total number of quarts of syrup produced that week?

5. One quart of maple syrup is sold for $18. How much would it cost to buy three quarts of syrup?

Sarah wanted to buy fabric for a craft project. After she figured out how many feet of each color she needed, Sarah discovered that fabric is sold by the yard. She made the following table to organize her information.

	feet	yard
red	2	
blue	3	
yellow	6	
white	5	
flowered	1	

6. How many feet are in one yard?

7. Should you multiply or divide by your answer to number 6 in order to change feet to yards?

8. Use your answers to numbers 6 and 7 to fill in the number of yards needed in the table. Some answers will be fractions or mixed numbers.

9. Most fabric is sold in fourths and eighths of a yard rather than thirds of a yard. Sarah needs $\frac{1}{3}$ of a yard of one of her fabrics. She knows that $\frac{1}{3}$ of a yard is more than $\frac{1}{4}$ of a yard. Will $\frac{3}{8}$ of a yard be enough to buy of that particular fabric?

10. If Sarah needs $\frac{2}{3}$ of a yard of fabric, should she ask the clerk for $\frac{1}{2}$ yard, $\frac{3}{8}$ yard, or $\frac{3}{4}$ of a yard? Choose the fraction that will give her what she needs with the least amount of extra fabric.

LESSON PRACTICE

Multiply the fractions. Use canceling as a shortcut. If you do not perform all possible cancellations, you will be able to simplify your answer further. Always be sure the answer is simplified as far as possible. The first two have been done for you.

1. $\dfrac{\cancel{2}^{1}}{4} \times \dfrac{\cancel{3}^{1}}{\cancel{6}_{3}} = \dfrac{1}{4}$
   _{1}

2. $\dfrac{5}{\cancel{6}_{3}} \times \dfrac{\cancel{3}^{1}}{7} \times \dfrac{\cancel{2}^{1}}{\cancel{3}_{1}} = \dfrac{5}{21}$

3. $\dfrac{1}{6} \times \dfrac{3}{8} = \underline{\hspace{1cm}}$

4. $\dfrac{1}{2} \times \dfrac{4}{5} = \underline{\hspace{1cm}}$

5. $\dfrac{1}{4} \times \dfrac{7}{11} \times \dfrac{4}{7} = \underline{\hspace{1cm}}$

6. $\dfrac{4}{5} \times \dfrac{1}{2} \times \dfrac{5}{8} = \underline{\hspace{1cm}}$

7. $\dfrac{1}{5} \times \dfrac{5}{6} = \underline{\hspace{1cm}}$

8. $\dfrac{1}{2} \times \dfrac{6}{7} \times \dfrac{2}{3} = \underline{\hspace{1cm}}$

9. $\dfrac{1}{4} \times \dfrac{3}{5} \times \dfrac{2}{3} = \underline{\hspace{1cm}}$

Change any mixed numbers to improper fractions and multiply. Use canceling as a shortcut. The first one has been done for you.

10. $\dfrac{6}{7} \times 1\dfrac{2}{5} = 1\dfrac{1}{5}$

$\dfrac{6}{\cancel{7}_{1}} \times \dfrac{\cancel{7}^{1}}{5} = \dfrac{6}{5} = 1\dfrac{1}{5}$

11. $2\dfrac{1}{5} \times 1\dfrac{3}{7} \times 2\dfrac{2}{3} =$ _____

12. $1\dfrac{5}{7} \times 1\dfrac{3}{4} =$ _____

13. $\dfrac{5}{6} \times 1\dfrac{2}{5} \times \dfrac{1}{7} =$ _____

14. There was $\frac{1}{2}$ of a pie in the refrigerator. Tom took $\frac{1}{2}$ of what was there and shared $\frac{1}{2}$ of his piece with Terry. What part of a pie did Terry get? $(\frac{1}{2} \times \frac{1}{2} \times \frac{1}{2} = ?)$

15. Three fourths of the snacks are Kitty's. She gives one half of her share to Susan, who gives two thirds of what she has to Michelle. What part of the total snacks does Michelle get? $(\frac{2}{3} \times \frac{1}{2} \times \frac{3}{4} = ?)$

It does not matter what order you use to multiply the fractions.

LESSON PRACTICE

25B

Multiply the fractions. Use canceling as a shortcut.

1. $\dfrac{4}{6} \times \dfrac{2}{5} =$ ——

2. $\dfrac{1}{6} \times \dfrac{3}{11} \times \dfrac{2}{3} =$ ——

3. $\dfrac{3}{8} \times \dfrac{2}{5} \times \dfrac{3}{3} =$ ——

4. $\dfrac{4}{5} \times \dfrac{2}{6} =$ ——

5. $\dfrac{2}{5} \times \dfrac{7}{8} \times \dfrac{4}{7} =$ ——

6. $\dfrac{1}{3} \times \dfrac{4}{9} \times \dfrac{3}{8} =$ ——

7. $\dfrac{1}{2} \times \dfrac{4}{6} =$ ——

8. $\dfrac{1}{3} \times \dfrac{6}{7} \times \dfrac{1}{5} =$ ——

9. $\dfrac{5}{8} \times \dfrac{1}{2} \times \dfrac{2}{5} =$ ——

Change any mixed numbers to improper fractions and multiply. Use canceling as a shortcut.

10. $4\frac{1}{6} \times 1\frac{3}{5} \times \frac{1}{10} =$ _____

11. $\frac{1}{2} \times \frac{4}{11} \times 1\frac{5}{6} =$ _____

12. $\frac{1}{6} \times 1\frac{3}{4} \times 1\frac{5}{7} =$ _____

13. $4\frac{1}{2} \times 1\frac{2}{3} \times 2\frac{1}{5} =$ _____

14. Ron had $2\frac{1}{2}$ bushels of apples, three fifths of which were spoiled. Ron added one third of the spoiled apples to his compost heap. What part of a bushel of apples was composted?

15. Forrest got $\frac{7}{10}$ of the total profit on the goods he sold. He gave $\frac{5}{6}$ of what he received to his wife. She gave $\frac{1}{7}$ of what she received to their son. What part of the total profit did the son receive? If the total profit for one job was $144, how much money did the son get?

Multiply the fractions. Use canceling as a shortcut.

1. $\dfrac{4}{5} \times \dfrac{1}{8}$ = _____

2. $\dfrac{2}{3} \times \dfrac{5}{6} \times \dfrac{3}{5}$ = _____

3. $\dfrac{3}{4} \times \dfrac{1}{5} \times \dfrac{5}{9}$ = _____

4. $\dfrac{7}{10} \times \dfrac{2}{21}$ = _____

5. $\dfrac{5}{9} \times \dfrac{5}{6} \times \dfrac{3}{5}$ = _____

6. $\dfrac{7}{8} \times \dfrac{2}{3} \times \dfrac{4}{5}$ = _____

7. $\dfrac{5}{6} \times \dfrac{3}{4}$ = _____

8. $\dfrac{3}{5} \times \dfrac{7}{10} \times \dfrac{4}{7}$ = _____

9. $\dfrac{5}{8} \times \dfrac{2}{3} \times \dfrac{4}{5}$ = _____

Change any mixed numbers to improper fractions and multiply. Use canceling as a shortcut.

10. $\dfrac{3}{8} \times 2\dfrac{2}{3} \times \dfrac{2}{3} =$ ———

11. $\dfrac{5}{6} \times \dfrac{2}{5} \times 1\dfrac{1}{2} =$ ———

12. $\dfrac{5}{9} \times 2\dfrac{4}{5} \times 2\dfrac{4}{7} =$ ———

13. $1\dfrac{1}{5} \times 3\dfrac{1}{3} \times 1\dfrac{3}{4} =$ ———

14. Mr. and Mrs. Wise are working on a family budget. One fifth of their income goes to savings; three eighths of that is for education, and five sixths of the education money is at Friendly Bank. What part of their total income is at Friendly Bank?

15. There were $4\dfrac{1}{2}$ pies left over after the Thanksgiving feast. Paula took one third of the leftover pies home. She gave Sandi one third of what she took home. What part of a pie did Sandi receive?

Change any mixed numbers to improper fractions and multiply. Use canceling as a shortcut.

1. $\dfrac{4}{7} \times 3\dfrac{1}{2} \times \dfrac{5}{6} =$ ——

2. $\dfrac{2}{9} \times \dfrac{3}{5} \times 2\dfrac{1}{2} =$ ——

Use the multiplicative inverse to solve each equation. Check your work.

3. $9G = 36$

4. Check for #3

5. $12V = 144$

6. Check for #5

Divide using the method you prefer. Use canceling when appropriate.

7. $5\dfrac{2}{5} \div 3\dfrac{2}{5} =$

8. $11\dfrac{5}{7} \div 1\dfrac{1}{7} =$

Add or subtract the mixed numbers.

9. $\begin{array}{r} 5\frac{1}{4} \\ -\ 2\frac{3}{4} \\ \hline \end{array}$

10. $\begin{array}{r} 7\frac{2}{5} \\ -\ 3\frac{7}{10} \\ \hline \end{array}$

11. $\begin{array}{r} 9\frac{1}{3} \\ +\ 6\frac{1}{4} \\ \hline \end{array}$

QUICK REVIEW

2,000 pounds (lb) = one ton

Fill in the blanks. Include a fraction in the answer if you cannot divide evenly. The first two have been done for you.

12. 5 tons = __10,000__ lb

   5 tons × 2,000 lb/ton = 10,000 lb

13. 7,000 lb = $3\frac{1}{2}$ tons

   7,000 lb ÷ 2,000 lb/ton = $3\frac{1}{2}$ tons

14. 3 tons = _____ lb

15. Five eighths of a large pizza had pepperoni topping. Sausage was added to one half of the part that had pepperoni, and peppers were added to one fifth of the pepperoni and sausage pizza. What part of the whole pizza had all three toppings?

16. Six times Jill's age is 300. How old is Jill? Write an equation and solve.

17. Emily bought $\frac{3}{4}$ of a ton of coal. How many pounds of coal does she have?

18. Mental Math! 40 minus 5, divided by 7, plus 3, times 7, equals _____ .

Convert any mixed numbers to improper fractions and multiply. Use canceling as a shortcut.

1. $\dfrac{1}{3} \times 1\dfrac{7}{8} \times \dfrac{4}{5} =$ _____

2. $2\dfrac{5}{6} \times 1\dfrac{1}{5} \times 3\dfrac{1}{8} =$ _____

Use the multiplicative inverse to solve each equation. Check your work.

3. $7H = 140$

4. Check for #3

5. $16W = 48$

6. Check for #5

Divide using the method you prefer.

7. $3\dfrac{1}{4} \div 4\dfrac{1}{5} =$

8. $\dfrac{4}{11} \div \dfrac{2}{11} =$

Add or subtract the mixed numbers.

9. $\begin{array}{r} 4\dfrac{1}{6} \\ -\ 1\dfrac{1}{2} \\ \hline \end{array}$

10. $\begin{array}{r} 8\dfrac{1}{2} \\ -\ 2\dfrac{3}{4} \\ \hline \end{array}$

11. $7\frac{4}{5}$

    $+\ 2\frac{1}{3}$
_____

Multiply the numerators and denominators by the same number to make equivalent fractions.

12. $\frac{2}{3}$ = _____ = _____ = $\frac{\ }{12}$

13. $\frac{1}{10}$ = _____ = $\frac{3}{\ }$ = _____

Fill in the blanks. Include a fraction in the answer if you cannot divide evenly.

14. 7 tons = _____ lb

15. 9,000 lb = _____ tons

16. $\frac{1}{2}$ ton = _____ lb

17. Carolyn and David bought $3\frac{1}{3}$ acres of land. One third of the land is for the house and yard, and three fifths of that part is for the backyard. How big is the backyard?

18. Keith bought $2\frac{1}{2}$ gallons of milk. How many quarts of milk did he buy?

19. Jasmine has $8\frac{3}{4}$ yards of ribbon. If she cuts the ribbon into equal pieces to give to five friends, how long is the piece that each person receives?

20. Mental Math! 60 plus 3, divided by 9, plus 4, times 3, equals _____ .

SYSTEMATIC REVIEW

25F

Change any mixed numbers to improper fractions and multiply. Use canceling as a shortcut.

1. $\dfrac{3}{5} \times 1\dfrac{1}{4} \times \dfrac{1}{4} =$ _____

2. $1\dfrac{7}{8} \times \dfrac{5}{6} \times 2\dfrac{2}{3} =$ _____

Use the multiplicative inverse to solve each equation. Check your work.

3. $6K = 54$

4. Check for #3

5. $25X = 450$

6. Check for #5

Divide using the method you prefer.

7. $1\dfrac{7}{9} \div 1\dfrac{2}{3} =$

8. $\dfrac{7}{8} \div \dfrac{1}{8} =$

Add or subtract the mixed numbers.

9. $10\dfrac{1}{3}$
    $- \ 3\dfrac{3}{4}$
    _____

10. $6\dfrac{2}{3}$
    $- \ 2\dfrac{4}{5}$
    _____

11.　　　$9\frac{4}{6}$

　　　$+\ 7\frac{5}{6}$

　　　‾‾‾‾‾‾‾‾

Multiply the numerators and denominators by the same number to make equivalent fractions.

12.　$\frac{4}{5}$ = ―――― = ―――― = $\frac{\ \ \ }{20}$　　　　13.　$\frac{3}{7}$ = ―――― = $\frac{9}{\ \ \ }$ = ――――

Fill in the blanks. Include a fraction in the answer if you are unable to divide evenly.

14.　4 tons = _____ lb　　　　　　　15.　7 pt = _____ qt

16.　$\frac{3}{4}$ lb = _____ oz

17.　Karen is $5\frac{1}{2}$ feet tall. How many inches tall is she?

18.　Karen is $5\frac{1}{2}$ feet tall. How many yards tall is she?

19.　A store advertised winter coats for $\frac{1}{3}$ of the original price. The next week the coats were being offered for $\frac{1}{2}$ of the sale price. After the heat wave, they were only $\frac{1}{4}$ of the last price. What part of the original price was the final price of the coats?

20.　Mental Math! 8 times 6, plus 2, divided by 5, times 10, equals _____ .

Multiplying a number by a fraction that is less than one will give an answer that is less than the number you started with. When multiplying by a mixed number, you are multiplying by a number greater than one, so you will get a result greater than the number you started with. Drawings can show how the answers to mixed-number multiplication problems make sense.

Mary's flower garden has an area of six square feet. She wants a garden that is two and one half times as large. Study the drawing. Since you are multiplying the area by a number greater than one, the new area will be larger than the original area.

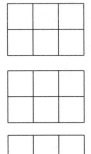

Each square represents one square foot. The first drawing shows the area of the original garden. Taken together, the drawings represent $2\frac{1}{2}$ times the original area. Notice that the squares could be arranged in any way that you wish and the area of the garden would not change.

1. Count the squares in the drawing to find the area of Mary's garden. Then multiply to see if you get the same answer.

There are $2\frac{1}{2}$ pizzas left over. Mom said that the boys could have one half of the leftover pizza for lunch. Study the drawings. In this case, the mixed number is greater than one, but we are multiplying by a fraction less than one. Therefore, the answer will be less than what we started with.

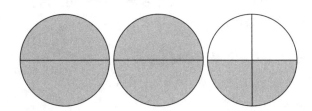

This drawing shows round pizzas.

2. How much pizza may the boys eat? Add together half of what is left in each drawing above to find the answer. Then multiply $\frac{1}{2}$ by $2\frac{1}{2}$ and compare the result to your first answer.

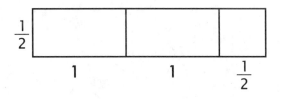

This drawing uses a rectangle to show the same problem. This is similar to how we build multiplication problems.

3. A rectangular plot of land measures $3\frac{1}{2}$ miles by $2\frac{1}{2}$ miles. What is its area? The problem is a mixed number times a mixed number. Both numbers are greater than one, and our answer will be greater than one.

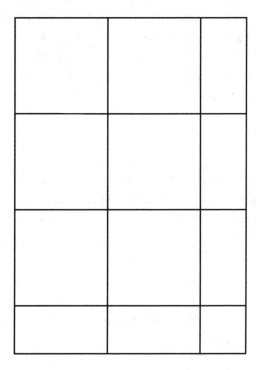

A rectangle like this is the most logical way to show an area problem. You may be able to read the units, halves, and quarters in the answer directly from the drawing. You may also write in the dimensions of each section, find the area of each one, and add them. Remember to confirm your answer by multiplying the mixed numbers.

Now try making your own drawings to solve these problems. Use another piece of paper if necessary. There may be more than one way to show each one. Remember to check your answer by multiplying the numbers.

4. A recipe calls for $1\frac{1}{2}$ cups of flour. Amanda wants to make three times the recipe. How many cups of flour does she need?

5. A garden is shaped like a square. It measures $2\frac{1}{2}$ yards on each side. What is the area of the garden?

Use multiplication, addition, and subtraction as needed to solve each equation. Check your work by replacing the unknown with the solution. The first two have been done for you.

1. 2X + 3 = 5

$$
\begin{array}{rcl}
2X + 3 &=& 5 \\
\underline{-3} & & \underline{-3} \\
2X + 0 &=& 2 \\
\dfrac{1}{2} \cdot 2X &=& 2 \cdot \dfrac{1}{2} \\
X &=& 1
\end{array}
$$

2. Check for #1

$$
\begin{array}{rcl}
2(1) + 3 &=& 5 \\
2 + 3 &=& 5 \\
5 &=& 5
\end{array}
$$

3. 4B + 8 = 28

4. Check for #3

5. 2Y – 11 = 7

6. Check for #5

7. 5C + 8 = 18

8. Check for #7

9. 5Z − 2 = 38

10. Check for #9

11. 3D + 9 = 21

12. Check for #11

13. 7A − 7 = 42

14. Check for #13

LESSON PRACTICE

Use multiplication, addition, and subtraction as needed to solve each equation. Check your work by replacing the unknown with the solution.

1. $5B + 3 = 18$

2. Check for #1

3. $12T - 1 = 35$

4. Check for #3

5. $4C + 7 = 31$

6. Check for #5

7. $6S - 13 = 11$

8. Check for #7

9. $10D + 11 = 91$

10. Check for #9

11. $7W - 13 = 22$

12. Check for #11

13. $3E + 17 = 20$

14. Check for #13

LESSON PRACTICE

Use multiplication, addition, and subtraction as needed to solve each equation. Check your work by replacing the unknown with the solution.

1. $5F + 13 = 38$

2. Check for #1

3. $8V - 14 = 26$

4. Check for #3

5. $7G + 21 = 56$

6. Check for #5

7. $5W - 12 = 53$

8. Check for #7

9. $6H + 7 = 25$

10. Check for #9

11. $11X - 16 = 72$

12. Check for #11

13. $2J + 17 = 37$

14. Check for #13

Use multiplication, addition, and subtraction as needed to solve each equation. Check your work by replacing the unknown with the solution.

1. $6Y - 12 = 12$

2. Check for #1

3. $4K + 6 = 46$

4. Check for #3

Change any mixed numbers to improper fractions and multiply. Use canceling as a shortcut.

5. $8\frac{1}{3} \times 6\frac{4}{5} \times \frac{1}{17} = $ _____

6. $\frac{3}{11} \times 2\frac{1}{5} \times 3\frac{1}{3} = $ _____

Divide using whichever method you prefer.

7. $\frac{5}{16} \div \frac{2}{8} = $

8. $4\frac{3}{5} \div 2\frac{1}{6} = $

Add.

9. $\frac{1}{2} + \frac{5}{6} + \frac{5}{9} = $ _____

10. $\frac{1}{4} + \frac{2}{3} + \frac{7}{12} = $ _____

QUICK REVIEW

5,280 feet (ft) = one mile (mi)

Fill in the blanks.

11. 2 miles = _____ ft

12. 7 mi = _____ ft

13. $\frac{1}{2}$ mi = _____ ft

14. A rectangle measures $3\frac{1}{2}$ feet by $6\frac{3}{4}$ feet. Find the area.

15. Is 3,518 divisible by 9?

16. Eight times Dan's age is 120. Write an equation and solve to find Dan's age.

17. What number is equivalent to 12^2?

18. Mental Math! 16 divided by 2, plus 2, times 3, divided by 6, equals _____ .

Use multiplication, addition, and subtraction as needed to solve each equation. Check your work by replacing the unknown with the solution.

1. $3Z - 13 = 20$

2. Check for #1

3. $8L + 7 = 55$

4. Check for #3

Change any mixed numbers to improper fractions and multiply. Use canceling as a shortcut.

5. $3\frac{1}{7} \times \frac{1}{2} \times \frac{5}{12} = $ ———

6. $\frac{5}{8} \times 3\frac{1}{3} \times 2\frac{1}{10} = $ ———

Divide using whichever method you prefer.

7. $\frac{1}{3} \div \frac{5}{18} = $

8. $5\frac{1}{3} \div 1\frac{1}{2} = $

Use the Rule of Four to make denominators the same. Then compare the fractions.

9. $\frac{3}{5}$ ◯ $\frac{1}{3}$

10. $\frac{4}{8}$ ◯ $\frac{2}{4}$

11. $\frac{1}{3}$ ◯ $\frac{4}{7}$

Fill in the blanks.

12. 5 miles = _____ ft

13. $\frac{2}{3}$ mi = _____ ft

14. $3\frac{1}{2}$ lb = _____ oz

15. Angie drove $25\frac{1}{3}$ miles. After she ran out of gas, she walked $1\frac{3}{4}$ miles. How far did Angie travel in all?

16. What is the greatest common factor (GCF) of 25 and 45?

17. Is 300 divisible by 5?

18. Kelly had $6\frac{1}{10}$ pounds of raisins. She divided them into bags for storage. Each bag could hold up to $1\frac{1}{2}$-pounds of raisins. How many bags did she use? (The last bag may be only partly full, but it still counts as another bag.)

19. What number is equivalent to 9^2?

20. Mental Math! 35 divided by 7, plus 4, times 6, plus 1, equals _____ .

Use multiplication, addition, and subtraction as needed to solve each equation. Check your work by replacing the unknown with the solution.

1. $7A + 8 = 29$

2. Check for #1

3. $9M - 9 = 54$

4. Check for #3

Change any mixed numbers to improper fractions and multiply. Use canceling as a short cut.

5. $1\frac{1}{5} \times 1\frac{1}{2} \times \frac{5}{6} =$ _____

6. $\frac{3}{7} \times \frac{5}{6} \times 2\frac{1}{3} =$ _____

Divide using whichever method you prefer.

7. $\frac{5}{8} \div \frac{1}{12} =$

8. $2\frac{2}{5} \div 1\frac{11}{25} =$

Subtract.

9. $\frac{3}{4} - \frac{1}{4} =$ _____

10. $\frac{1}{2} - \frac{1}{5} =$ _____

11. $\frac{4}{5} - \frac{1}{6} =$ _____

Fill in the blanks. Include a fraction in the answer if you are unable to divide evenly.

12. 35 ft = _____ yd

13. 8 qt = _____ pt

14. $1\frac{1}{2}$ gal = _____ qt

15. A zookeeper ordered $10\frac{1}{2}$ tons of hay for his animals. If $6\frac{3}{4}$ tons have been eaten, how many *pounds* of hay are left?

16. Find the prime factors of 63.

17. If Stan is $6\frac{1}{2}$ feet tall, how many inches tall is he?

18. Sue has to walk $\frac{7}{8}$ of a mile to school. How many feet must she walk? How many yards must Sue walk to school?

19. What number is equivalent to 23^2?

20. Mental Math! 42 divided by 6, minus 1, times 6, plus 4, equals _____ .

Division problems with mixed numbers may be represented in different ways.

1. Mary had $3\frac{1}{2}$ yards of fabric. She divided it into pieces that were each $\frac{1}{4}$ of a yard long. How many pieces did she have when she was finished?

   The drawing represents the length of fabric. The width is not important in this case. Draw lines to show how Mary cut the fabric into lengths that are each $\frac{1}{4}$ yard long. Check your answer by division.

2. There are $2\frac{3}{4}$ pies left on the counter. If each person has a piece that is $\frac{1}{8}$ of a whole pie, how many people can have pie?

   The drawing represents the pies after they have been cut into eighths.

   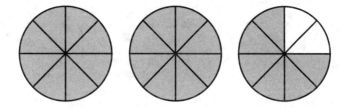

For some types of mixed number division problems, it is helpful to make a drawing after you know the answer. You may use the drawing as a way to check your answer.

3. A rectangular window has an area of $3\frac{3}{4}$ square feet. The width of the window is $1\frac{1}{2}$ feet. What is the height of the window? Review number 3 on Application and Enrichment 25G. See if you can make a drawing that illustrates this problem. If you prefer, you can do the division first and then make the drawing once you know both dimensions of the window.

4. Rachel has five cups of fresh blueberries. How many times can she make a muffin recipe that calls for $1\frac{1}{4}$ cups of berries?

Draw lines to divide the cups into fourths.

Use different colors or shading to show how the berries can be divided into parts that are each $1\frac{1}{4}$ cups.

Divide to confirm your answer. $5 \div 1\frac{1}{4} = $ _____

5. Seven and one half big chocolate chip cookies are to be divided evenly among three people. How many cookies should each person receive?

Draw lines around the groups of whole cookies that each person will receive. Then divide the leftovers evenly among three people. The result should add up to the same number that you get by dividing.

6. A trail is $6\frac{2}{3}$ miles long. A bench is placed every $1\frac{1}{3}$ miles. Into how many sections has the trail been divided? Solve using division and then make a drawing to illustrate the problem. If there are benches at the beginning and end of the trail, how many benches are there altogether?

Find the area of each circle. Use $\frac{22}{7}$ as an approximate value of π. The first one has been done for you.

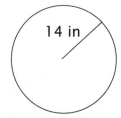

1. A ≈ <u>616 sq in</u>

    $A = \pi r^2$

    $(14 \text{ in})^2 = 196 \text{ sq in}$

    $$\frac{22}{\cancel{7}} \times \frac{\overset{28}{\cancel{196}} \text{ sq in}}{1} = \frac{616}{1} = 616 \text{ sq in}$$

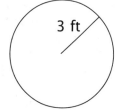

2. A ≈ _____

3. A ≈ _____

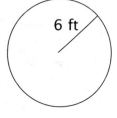

4. A ≈ _____

5. A pond is circular with a radius of four yards. What is the approximate area of the surface of the water?

Find the circumference of each circle. Use $\frac{22}{7}$ as an approximate value of π. The first one has been done for you.

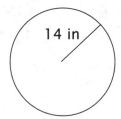

6. C ≈ <u>88 in</u>

C = 2πr

$$\frac{2}{1} \times \frac{22}{\cancel{7}} \times \frac{\overset{2}{\cancel{14}} \text{ in}}{1} = 88 \text{ in}$$

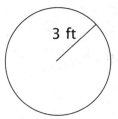

7. C ≈ _____

8. C ≈ _____

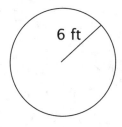

9. C ≈ _____

10. Kate's new dinner plates have a radius of seven inches. What is the approximate circumference of each plate?

LESSON PRACTICE

Find the area of each circle. Use $\frac{22}{7}$ as an approximate value of π.

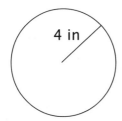

1. A ≈ _____

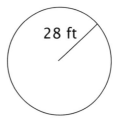

2. A ≈ _____

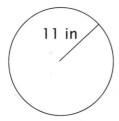

3. A ≈ _____

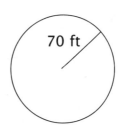

4. A ≈ _____

5. The cowboys tethered their horses after the roundup. Each horse had a ten-foot rope. What was the approximate area of the circle that each horse had for grazing?

Find the circumference of each circle. Use $\frac{22}{7}$ as an approximate value of π.

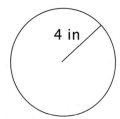

6. C ≈ _____

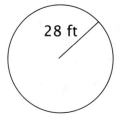

7. C ≈ _____

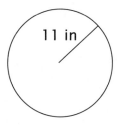

8. C ≈ _____

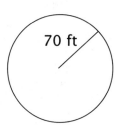

9. C ≈ _____

10. The paper Stephanie is using for her art project measures nine inches from the center to the edge. What would be the approximate circumference of the biggest circle she could cut from it?

Find the area of each circle. Use $\frac{22}{7}$ as an approximate value of π.

1. A ≈ _____

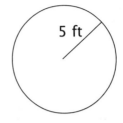

2. A ≈ _____

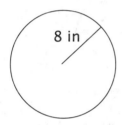

3. A ≈ _____

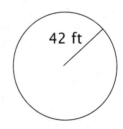

4. A ≈ _____

5. Alison plans to paint the bottom of her round swimming pool. She needs to know the area so she can figure out how much paint to buy. The radius of the pool is 12 feet. What is the approximate area of the pool bottom?

Find the circumference of the circles. Use $\frac{22}{7}$ as an approximate value of π.

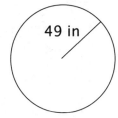

6. C ≈ _____

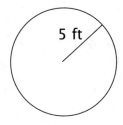

7. C ≈ _____

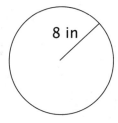

8. C ≈ _____

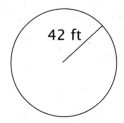

9. C ≈ _____

10. An airplane repeatedly flew around a circle with a radius of 35 miles. Approximately how many miles did the airplane travel in one flight around the circle? (circumference)

SYSTEMATIC REVIEW

Find the area and circumference of each circle. Use $\frac{22}{7}$ as an approximate value of π.

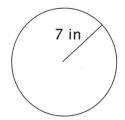

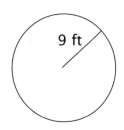

1. A ≈ _____

2. C ≈ _____

3. A ≈ _____

4. C ≈ _____

Use the multiplicative inverse to solve each equation. Check your work.

5. $3P - 15 = 12$

6. Check for #5

Change any mixed numbers to improper fractions and multiply.

7. $1\frac{1}{5} \times \frac{5}{12} \times \frac{3}{7} =$ _____

8. $\frac{3}{5} \times 1\frac{6}{14} \times 3\frac{1}{2} =$ _____

Divide using the method you prefer.

9. $\frac{9}{11} \div \frac{1}{11} =$

10. $2\frac{1}{4} \div 1\frac{5}{22} =$

QUICK REVIEW

The same symbols used to compare fractions can also be used to compare measures.

Use <, >, or = to compare the measures. The first one has been done for you.

11. 10 ft (>) 3 yd

10 ft (>) 9 ft

12. 8 pt ( ) 4 qt

13. $\frac{1}{2}$ lb ( ) 7 oz

14. What is the approximate area of floor covered by a circular rug that has a radius of 16 inches?

15. Each side of a cube measures one half of a foot. What is the volume of the cube?

16. Change the length of the sides of the cube in #15 to inches and find the volume in cubic inches.

17. An aphid crawled $1\frac{1}{2}$ inches, $2\frac{1}{4}$ inches, and $5\frac{1}{4}$ inches. How far did the aphid crawl?

18. Mental Math! 10 times 10, divided by 2, plus 3, minus 4, equals _____ .

SYSTEMATIC REVIEW

Find the area and circumference of each circle. Use $\frac{22}{7}$ as an approximate value of π.

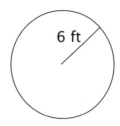

1. A ≈ _____

3. A ≈ _____

2. C ≈ _____

4. C ≈ _____

Use the multiplicative inverse to solve each equation. Check your work.

5. 9D + 14 = 23

6. Check for #5

Change any mixed numbers to improper fractions and multiply.

7. $2\frac{1}{4} \times 2\frac{2}{3} \times \frac{2}{9} =$ _____

8. $\frac{5}{6} \times 4\frac{1}{2} \times 1\frac{1}{3} =$ _____

Divide using the method you prefer.

9. $\frac{4}{7} \div 2\frac{1}{3} =$

10. $10\frac{2}{3} \div 4\frac{1}{4} =$

Use <, >, or = to compare the measures.

11. 15 qt $\bigcirc$ 8 gal

12. 2 tons $\bigcirc$ 1,500 lb

13. $\frac{1}{3}$ mi $\bigcirc$ 1,760 ft

14. A $\frac{3}{4}$-ton pickup truck does not necessarily weigh $\frac{3}{4}$ tons; it can haul $\frac{3}{4}$ tons. How many pounds can it haul?

15. Kimberly has a round flower bed with a radius of 14 feet. What is the approximate area of her flower bed?

16. Kimberly wants to put a fence around her flower bed. (See #15.) About how many feet of fencing must she buy? (circumference)

17. The base of a triangle is $4\frac{1}{2}$ inches, and the height is 8 inches. What is the area of the triangle?

18. It takes $1\frac{1}{6}$ dozen eggs to make a fancy angel food cake. If Esther has $10\frac{1}{2}$ dozen eggs, how many cakes can she make?

19. Three times Anthony's age is 90. Write an equation and solve to find Anthony's age.

20. Mental Math! 11 times 3, minus 1, divided by 8, times 7, equals _____ .

SYSTEMATIC REVIEW

Find the area and circumference of each circle. Use $\frac{22}{7}$ as an approximate value of π.

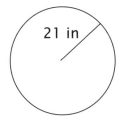

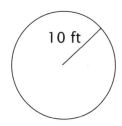

1. A ≈ _____

3. A ≈ _____

2. C ≈ _____

4. C ≈ _____

Use the multiplicative inverse to solve each equation. Check your work.

5. 4Q – 5 = 3

6. Check for #5

Change any mixed numbers to improper fractions and multiply.

7. $2\frac{2}{5} \times 2\frac{3}{4} =$ _____

8. $5\frac{5}{8} \times 1\frac{3}{5} \times 2\frac{1}{7} =$ _____

Divide using the method you prefer.

9. $\frac{4}{10} \div \frac{2}{5} =$

10. $9\frac{3}{7} \div 2\frac{1}{10} =$

Use <, >, or = to compare the measures.

11. 17 ft $\bigcirc$ 17 in

12. 5 tons $\bigcirc$ 10,000 lb

13. $\frac{3}{4}$ mi $\bigcirc$ 4,000 ft

14. A town is planning a circular park with a radius of one mile. What will the approximate area of the park be?

15. The mayor wants the park (#14) to have an approximate radius of two miles. What would the new area of the park be? Would doubling the radius make the area of the park two times as great?

16. Paula bought 10 pounds of flour. She used $6\frac{3}{4}$ pounds to make bread. How many pounds of flour are left over?

17. The base of a triangle is 9 inches long, and the height is $3\frac{1}{3}$ inches. What is the area of the triangle?

18. Dave's rectangular backyard measures $20\frac{1}{3}$ feet by $30\frac{1}{2}$ feet. What is the perimeter of his yard?

19. Dave wants to put a fence around $\frac{3}{5}$ of the perimeter of his yard. (See #18.) How many feet of fencing should he buy?

20. Mental Math! 12 plus 8, minus 10, times 4, plus 7, equals _____ .

There are many ways to describe patterns. This pattern was built using squares and triangles. Study the design. Draw more steps if necessary and then fill in the chart. Use the chart to help you answer the questions.

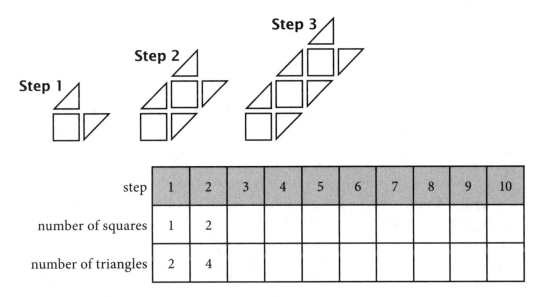

step	1	2	3	4	5	6	7	8	9	10
number of squares	1	2								
number of triangles	2	4								

1. How many squares are needed to build the fifth design in the sequence?

2. How many triangles are needed to build the fifth design in the sequence?

3. How would you describe the pattern of the numbers in the chart?

4. How many of each shape would be needed to build the twentieth design in the sequence?

5. Frank the Frog is chasing flies. For every one he catches, three get away. The numbers on the chart tell how many flies he has caught or missed since he started recording the data. Follow the pattern to complete the chart.

total flies caught	1	2							
total flies missed	3	6							

6. Use the chart to tell how many flies Frank will have missed when he has caught five flies.

7. Use the chart to tell how many flies Frank will have caught when he has missed 27 flies.

8. Whenever Paul went to the cookie jar, he took out three cookies. One he ate right away, and two he saved to eat later. Follow the pattern to complete the chart.

visits to the cookie jar	1	2	3						
total cookies eaten	1	2							
total cookies saved	2	4							

9. How many cookies were saved after six visits to the cookie jar?

10. After seven trips to the cookie jar, how many cookies has Paul taken out of the jar altogether?

11. If Paul has saved 40 cookies, how many did he eat right away?

LESSON PRACTICE

Use addition and subtraction as needed. Then use the reciprocal or multiplicative inverse of the coefficient to solve each equation. Check your work. The first two have been done for you.

1. $\frac{5}{6}X + 3 = 13$

$$\frac{5}{6}X + 3 = 13$$
$$\quad\ \ \underline{-3\quad -3}$$
$$\frac{5}{6}X + 0 = 10$$
$$\frac{6}{5}\cdot\frac{5}{6}X \ = \overset{2}{\cancel{10}}\cdot\frac{6}{\cancel{5}}$$
$$X\quad = 12$$

2. Check for #1
$$\frac{5}{6}(12) + 3 = 13$$
$$10 + 3 = 13$$
$$13 = 13$$

3. $\frac{1}{5}A = 3$

4. Check for #3

5. $\frac{1}{7}Y - 4 = 2$

6. Check for #5

7. $\frac{3}{8}B + 8 = 14$

8. Check for #7

9. $\dfrac{2}{3}Z = 12$

10. Check for #9

11. $\dfrac{2}{5}C - 2 = 2$

12. Check for #11

Use multiplication, addition, and subtraction as needed to solve each equation. Check your work by replacing the unknown with the solution.

1. $\dfrac{1}{4}G - 4 = 2$

2. Check for #1

3. $\dfrac{4}{5}D = 40$

4. Check for #3

5. $\dfrac{1}{10}H - 4 = 0$

6. Check for #5

7. $\dfrac{5}{8}E - 5 = 20$

8. Check for #7

9. $\dfrac{1}{4} J = 7$

10. Check for #9

11. $\dfrac{5}{9} F + 9 = 14$

12. Check for #11

Use multiplication, addition, and subtraction as needed to solve each equation. Check your work by replacing the unknown with the solution.

1. $\frac{1}{3}K + 3 = 11$

2. Check for #1

3. $\frac{1}{5}G = 8$

4. Check for #3

5. $\frac{3}{5}L + 3 = 18$

6. Check for #5

7. $\frac{5}{12}R + 5 = 25$

8. Check for #7

9. $\dfrac{2}{3} M = 10$

10. Check for #9

11. $\dfrac{2}{9} S + 6 = 14$

12. Check for #11

SYSTEMATIC REVIEW

Use multiplication, addition, and subtraction as needed to solve each equation. Check your work by replacing the unknown with the solution.

1. $\frac{1}{5}T + 4 = 8$

2. Check for #1

3. $\frac{3}{4}N - 6 = 21$

4. Check for #3

Find the area and circumference of each circle. Use $\frac{22}{7}$ as an approximate value of π.

14 ft

5. A ≈ _____

6. C ≈ _____

Change any mixed numbers to improper fractions and multiply.

7. $\frac{1}{2} \times \frac{2}{3} \times \frac{9}{11} =$ _____

8. $\frac{7}{8} \times 2\frac{2}{3} \times \frac{6}{7} =$ _____

Divide using whichever method you prefer.

9. $\dfrac{3}{1} \div \dfrac{1}{2} =$ _____

10. $6\dfrac{1}{5} \div 1\dfrac{11}{20} =$ _____

Use <, >, or = to compare the measures.

11. 21 ft ⬭ 12 yd

12. 5 qt ⬭ 11 pt

13. 5 qt ⬭ 1 gal

QUICK REVIEW

To find the average of a series of numbers, add the numbers and divide the answer by the number of items in the list.

Example 1
Find the average of 3, 4, and 5.
$3 + 4 + 5 = 12$ and $12 \div 3 = 4$.

Find the average of each series of numbers.

14. 2, 7, 9   Average = _____

15. 5, 5, 13, 9   Average = _____

16. 2, 8, 5, 9   Average = _____

17. Jim earned the following scores on math tests last month: 95, 80, 77, 100. What was his average?

18. Mental Math! 2 plus 7, times 7, plus 2, plus 5, equals _____ .

Use multiplication, addition, and subtraction as needed to solve each equation. Check your work by replacing the unknown with the solution.

1. $\frac{1}{6}U + 2 = 3$

2. Check for #1

3. $\frac{2}{5}P = 2$

4. Check for #3

Find the area and circumference of each circle. Use $\frac{22}{7}$ as an approximate value of π.

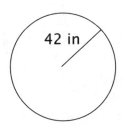

42 in

5. $A \approx$ _____

6. $C \approx$ _____

Change any mixed numbers to improper fractions and multiply.

7. $1\frac{1}{14} \times 1\frac{3}{4} \times \frac{1}{25} =$ _____

8. $\frac{4}{9} \times 2\frac{1}{2} \times \frac{3}{5} =$ _____

Divide using the method you prefer.

9. $\dfrac{3}{6} \div \dfrac{2}{3} =$ _____

10. $3\dfrac{3}{7} \div 2\dfrac{4}{7} =$ _____

Use <, >, or = to compare the measures.

11. 32 oz $\bigcirc$ 2 lb

12. 7 tons $\bigcirc$ 8,000 lb

13. 10 ft $\bigcirc$ 125 in

Find the average of each series of numbers.

14. 5, 4, 6   Average = _____

15. 10, 7, 6, 9   Average = _____

16. 7, 4, 10   Average = _____

17. The dimensions of a rectangle are $4\frac{1}{2}$ inches and $6\frac{1}{3}$ inches. What is the perimeter of the rectangle?

18. What is the area of the rectangle in #17?

19. What are the prime factors of 64?

20. Mental Math! 3 times 2, times 3, plus 2, minus 4, equals _____ .

Use multiplication, addition, and subtraction as needed to solve each equation. Check your work by replacing the unknown with the solution.

1. $\frac{3}{16} V - 3 = 12$

2. Check for #1

3. $5Q + 4 = 19$

4. Check for #3

Find the area and circumference of each circle. Use $\frac{22}{7}$ as an approximate value of π.

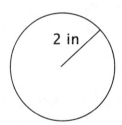

2 in

5. $A \approx$ _____

6. $C \approx$ _____

Change any mixed numbers to improper fractions and multiply.

7. $2 \times \frac{3}{10} \times \frac{5}{6} =$ _____

8. $\frac{10}{21} \times 2\frac{4}{5} \times \frac{6}{11} =$ _____

Divide using the method you prefer.

9. $\dfrac{1}{2} \div \dfrac{1}{10} =$ _____

10. $4\dfrac{5}{8} \div 3\dfrac{5}{12} =$ _____

Use <, >, or = to compare the measures.

11. 16 lb $\bigcirc$ 1 oz

12. $\dfrac{1}{2}$ mi $\bigcirc$ 2,640 ft

13. $\dfrac{3}{4}$ ft $\bigcirc$ 10 in

Find the average of each series of numbers.

14. 4, 8, 10, 14 Average = _____

15. 6, 9, 7, 14 Average = _____

16. 6, 6, 12 Average = _____

17. Isaac ran $1\dfrac{1}{2}$ miles the first day, $2\dfrac{1}{4}$ miles the second day, and $1\dfrac{1}{4}$ miles the third day. What is the average number of miles he ran per day? Include a fraction in the answer if you cannot divide evenly.

18. A cube measures $1\dfrac{1}{3}$ inches on each side. What is the volume of the cube?

19. What is the GCF of 30 and 45?

20. Mental Math! 20 divided by 5, times 3, plus 7, minus 4, equals _____ .

You have learned to multiply mixed numbers by changing them to improper fractions. Some students have asked why they can't multiply vertically, since addition and subtraction of mixed numbers is done vertically.

As a matter of fact, you can multiply mixed numbers vertically. After you try a few, you will understand why it is easier to do these problems with improper fractions. As you study the solution, keep in mind what you know about place value and multiplication.

$$2\;\frac{1}{4}$$

$$\times 1\;\frac{2}{5}$$

$$\frac{4}{5}\quad\frac{2}{20}$$

$\frac{2}{5} \times \frac{1}{4}$ is $\frac{2}{20}$. Put the answer in a column to the right of the fractions. $\frac{2}{5} \times 2$ is $\frac{4}{5}$. Put the answer in the fraction place. $1 \times \frac{1}{4}$ is $\frac{1}{4}$. Put the answer in the fraction place. 1×2 is 2. Put the answer in the units place.

$$2\;\frac{1}{4}$$

Vertically add the partial products, using the rule of 4 for the middle terms.

$$2\;\frac{21}{20}+\frac{2}{20}$$

Add $\frac{21}{20}$ and $\frac{2}{20}$.

$$2+\frac{23}{20}$$

The result is similar to expanded notation.
Change the improper fraction to a mixed number.

$$2+1\frac{3}{20}=3\frac{3}{20}$$

Add to get the final answer.

When you convert mixed numbers to improper fractions, the solution looks like this:

$$2\frac{1}{4}\times 1\frac{2}{5}=\frac{9}{4}\times\frac{7}{5}=\frac{63}{20}=3\frac{3}{20}$$

Solve each problem vertically and then by using improper fractions. As you work through each problem, be sure to remember when you are multiplying the factors and when you are adding the partial products.

Final answers may be whole numbers, mixed numbers, or fractions. Be sure to simplify any answers that include fractions.

1. $\begin{array}{r} 4\frac{1}{2} \\ \times\, 5\frac{1}{3} \\ \hline \end{array}$

2. $\begin{array}{r} 3\frac{1}{7} \\ \times\, 2\frac{1}{2} \\ \hline \end{array}$

Build each fraction and then use both tenths overlays to change the fraction to a decimal. Shade the drawings to show tenths and hundredths. The first one has been done for you.

1.

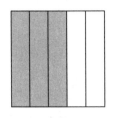 →  →

$$\frac{3}{5}$$ $$\frac{6}{10} = \underline{0.6}$$ $$\frac{60}{100} = \underline{0.60}$$

2.

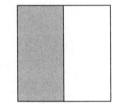 → 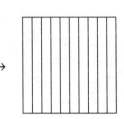 →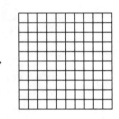

$$\frac{1}{2}$$ $$\frac{}{10} = \underline{}$$ $$\frac{}{100} = \underline{}$$

3.

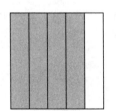 → 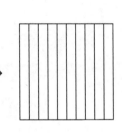 →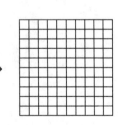

$$\frac{4}{5}$$ $$\frac{}{10} = \underline{}$$ $$\frac{}{100} = \underline{}$$

Change each fraction to an equivalent fraction in hundredths. Then change it to a percent. You may use the overlays to help you change the fractions to hundredths. The first one has been done for you.

4. $\dfrac{1}{5} = \dfrac{20}{100} = 20\%$

5. $\dfrac{5}{5} = \dfrac{}{100} = \underline{}\%$

Write each fraction in hundredths. Then write it as a decimal and as a percent. The first one has been done for you.

6. $\dfrac{2}{5} = \dfrac{40}{100} = 0.40 = 40\%$

7. $\dfrac{1}{2} = \dfrac{}{100} = \underline{} = \underline{}\%$

Build each fraction and then use both tenths overlays to change the fraction to a decimal. Shade the drawings to show hundredths. The first one has been done for you.

1.

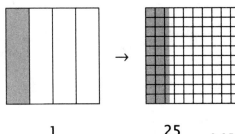

$$\frac{1}{4}$$ $$\frac{25}{100} = \underline{0.25}$$

2.

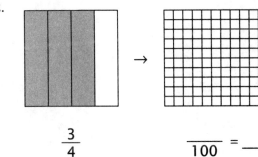

$$\frac{3}{4}$$ $$\overline{100} = \underline{}$$

Build each fraction and then use both tenths overlays to change the fraction to a decimal. Shade the drawings to show tenths and hundredths.

3.

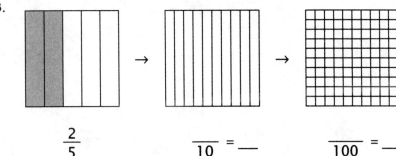

$$\frac{2}{5}$$ $$\overline{10} = \underline{}$$ $$\overline{100} = \underline{}$$

Change each fraction to an equivalent fraction in hundredths and then to a percent. You may use the overlays to help you change the fractions to hundredths.

4. $\dfrac{3}{5}$ = $\dfrac{}{100}$ = _____ %

5. $\dfrac{4}{5}$ = $\dfrac{}{100}$ = _____ %

Write each fraction in hundredths. Then write it as a decimal and as a percent.

6. $\dfrac{1}{4}$ = $\dfrac{}{100}$ = _____ = _____ %

7. $\dfrac{3}{4}$ = $\dfrac{}{100}$ = _____ = _____ %

LESSON PRACTICE

Build each fraction and then use both tenths overlays to change the fraction to a decimal. Shade the drawings to show tenths or hundredths.

1.

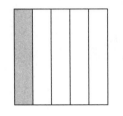 → 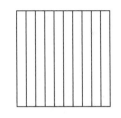 →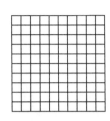

$\dfrac{1}{5}$ $\dfrac{}{10} = \dfrac{}{}$ $\dfrac{}{100} = \dfrac{}{}$

2.

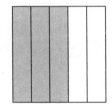 →  →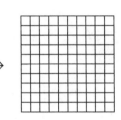

$\dfrac{3}{5}$ $\dfrac{}{10} = \dfrac{}{}$ $\dfrac{}{100} = \dfrac{}{}$

3.

 →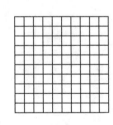

$\dfrac{1}{4}$ $\dfrac{}{100} = \dfrac{}{}$

Change each fraction to an equivalent fraction in hundredths and then to a percent. You may use the overlays to help you change the fractions to hundredths.

4. $\dfrac{5}{5} = \dfrac{}{100} = $ _____ %

5. $\dfrac{2}{5} = \dfrac{}{100} = $ _____ %

Write each fraction in hundredths. Then write it as a decimal and as a percent.

6. $\dfrac{3}{4} = \dfrac{}{100} = $ _____ = _____ %

7. $\dfrac{1}{2} = \dfrac{}{100} = $ _____ = _____ %

Change each fraction to an equivalent fraction in hundredths and then to a percent. You may use the overlays to help you change the fractions to hundredths.

1. $\dfrac{1}{5} = \dfrac{\rule{2em}{0.4pt}}{100} = \rule{3em}{0.4pt}\,\%$

2. $\dfrac{3}{5} = \dfrac{\rule{2em}{0.4pt}}{100} = \rule{3em}{0.4pt}\,\%$

Write each fraction in hundredths. Then write it as a decimal and as a percent.

3. $\dfrac{1}{4} = \dfrac{\rule{2em}{0.4pt}}{100} = \rule{3em}{0.4pt} = \rule{3em}{0.4pt}\,\%$

4. $\dfrac{4}{5} = \dfrac{\rule{2em}{0.4pt}}{100} = \rule{3em}{0.4pt} = \rule{3em}{0.4pt}\,\%$

Solve and check your work.

5. $\dfrac{1}{2}T + 5 = 8$

6. Check for #5

Change any mixed numbers to improper fractions and multiply.

7. $\dfrac{3}{4} \times \dfrac{2}{7} \times \dfrac{1}{3} = \rule{3em}{0.4pt}$

8. $5\dfrac{2}{8} \times 7\dfrac{5}{6} \times 1\dfrac{5}{7} = \rule{3em}{0.4pt}$

QUICK REVIEW

The numbers we use every day are called Arabic numerals. Sometimes you will see letters used to indicate numbers. These are called Roman numerals. In Roman numerals, I stands for 1, V stands for 5, and X stands for 10. I and X may be used up to three times in a row, but V may be used only once.

I written before another letter indicates subtraction. Study the chart.

1 I	8 VIII	15 XV	25 XXV
2 II	9 IX	16 XVI	30 XXX
3 III	10 X	17 XVII	35 XXXV
4 IV	11 XI	18 XVIII	39 XXXIX
5 V	12 XII	19 XIX	
6 VI	13 XIII	20 XX	
7 VII	14 XIV	21 XXI	

Write the Arabic numeral that corresponds to each Roman numeral.

9. VI _____

10. XIII _____

11. IX _____

12. XXXI _____

Write the Roman numeral that corresponds to each Arabic numeral.

13. 20 _____

14. 8 _____

15. 16 _____

16. 34 _____

17. The ages of three sisters are 16, 18, and 22. What is their average age? (Include a fraction in the answer if you are unable to divide evenly.)

18. What are the approximate area and circumference of a circle that has a radius of 70 feet?

Change each fraction to an equivalent fraction in hundredths and then to a percent. You may use the overlays to help you change the fractions to hundredths.

1. $\dfrac{2}{5} = \dfrac{}{100} = $ _____ %

2. $\dfrac{5}{5} = \dfrac{}{100} = $ _____ %

Write each fraction in hundredths. Then write it as a decimal and as a percent.

3. $\dfrac{3}{4} = \dfrac{}{100} = $ _____ $= $ _____ %

4. $\dfrac{1}{2} = \dfrac{}{100} = $ _____ $= $ _____ %

Solve and check your work.

5. $\dfrac{4}{5}U - 11 = 33$

6. Check for #5

Add the fractions.

7. $\dfrac{3}{4} + \dfrac{2}{3} + \dfrac{1}{12} = $ _____

8. $\dfrac{1}{3} + \dfrac{2}{5} + \dfrac{1}{15} = $ _____

Write the Arabic numeral that corresponds to each Roman numeral.

9. II _____

10. XVII _____

11. XXVII _____

12. XIX _____

Write the Roman numeral that corresponds to each Arabic numeral.

13. 7 _____   14. 18 _____

15. 21 _____   16. 35 _____

17. The base of a triangle is 24 inches, and the height is $6\frac{1}{2}$ inches. What is the area of the triangle?

18. The area of a rectangle is $\frac{3}{8}$ of a square mile. One dimension is $\frac{5}{8}$ of a mile. What is the length of the other dimension of the rectangle?

19. Three lines have the following lengths: $3\frac{1}{3}$ inches, $6\frac{1}{2}$ inches, and $5\frac{1}{6}$ inches. What is the average length of the lines?

20. Sonia planted a circular garden with a radius of seven feet. Approximately how many feet of fence are needed to go around the edge of her garden?

Change each fraction to an equivalent fraction in hundredths and then to a percent.

1. $\dfrac{3}{5} = \dfrac{}{100} = $ _____ %

2. $\dfrac{1}{5} = \dfrac{}{100} = $ _____ %

Write each fraction in hundredths. Then write it as a decimal and as a percent.

3. $\dfrac{4}{5} = \dfrac{}{100} = $ _____ $= $ _____ %

4. $\dfrac{1}{4} = \dfrac{}{100} = $ _____ $= $ _____ %

Solve and check your work.

5. $7X - 9 = 19$

6. Check for #5

Add or subtract the mixed numbers.

7. $16\dfrac{5}{9}$
$- 8\dfrac{1}{9}$

8. 7
$- 3\dfrac{4}{5}$

9. $4\dfrac{3}{8}$
$+ 2\dfrac{3}{5}$

Write the Arabic numeral that corresponds to each Roman numeral.

10. III _____

11. XI _____

12. XXV _____

13. XXXIX _____

Write the Roman numeral that corresponds to each Arabic numeral.

14. 5 _____

15. 12 _____

16. 38 _____

17. 14 _____

18. Greg traveled 45 miles on Monday, 34 miles on Tuesday, and 40 miles on Wednesday. What is the average distance he traveled each day?

19. The boss said that $\frac{1}{2}$ of the job should be finished this week. He asked Russ to do $\frac{1}{4}$ of this week's part of the job. Wayne was not busy, so he helped Russ by doing $\frac{4}{5}$ of what the boss asked Russ to do. What part of the total job did Wayne do?

20. Keith has a circular lawn with a radius of 28 feet. One bag of grass seed covers 200 square feet. Will one bag be enough for Keith's lawn?

APPLICATION AND ENRICHMENT

You have learned how to compare fractions by making the denominators the same. One advantage of using decimals is that they are easy to compare. Notice that there is a zero before the decimal in numbers that have no unit place value. This makes it easier to notice the decimal point and not confuse 0.4 with 4.

Here are some examples of comparing decimals.

Example 1

Which decimal is greater: 0.4 or 0.5?

Both decimals are expressed as tenths, and since 5 is greater than 4, the answer is 0.5. You can write the relationship as 0.5 > 0.4.

Example 2

Which number is less: 2.32 or 2.56?

The number in the units place is the same for both numbers. Look at the decimal parts of the numbers. Both decimals are expressed as hundredths, and 32 is less than 56. The answer is 2.32.

You can write the relationship as 2.32 < 2.56.

Example 3

Compare the decimals: 0.8 and 0.80

You can compare only like to like. The decimal 0.80 can be rewritten as 0.8 without changing its value. You can see why this is true if you rewrite the decimals as $\frac{8}{10}$ and $\frac{80}{100}$. Since $\frac{80}{100}$ can be simplified to $\frac{8}{10}$, the fractions are equivalent.

$0.8 = 0.80$

Example 4

Compare the numbers: 0.543 and 1.432

The numbers in the tenths, hundredths, and thousandths places are all greater in the first number given. However, if you look at the units place, you see that 1 is greater than 0, so the second number given is actually greater. Always compare the numbers in the greatest place value position first.

You can write the answer as 0.543 < 1.432 or as 1.432 > 0.543.

Here are some exercises to help you practice comparing decimals. Use the strategies from the previous page. Write <, >, or = in each oval.

1. 4.2 $\bigcirc$ 3.8

2. 0.90 $\bigcirc$ 0.9

3. 2.31 $\bigcirc$ 1.31

4. 0.57 $\bigcirc$ 0.75

5. 0.123 $\bigcirc$ 0.238

6. 0.8 $\bigcirc$ 0.12

7. 1.62 $\bigcirc$ 0.83

Use multiplication, addition, and subtraction as needed to solve each equation. Check your work by replacing the unknown with the solution. The first two have been done for you.

1. $\dfrac{1}{4}K - \dfrac{1}{2} = \dfrac{1}{4}$

$$\dfrac{1}{4}K - \dfrac{1}{2} = \dfrac{1}{4}$$
$$\quad\quad + \dfrac{1}{2} \quad + \dfrac{1}{2}$$
$$\dfrac{4}{1} \cdot \dfrac{1}{4}K = \dfrac{3}{4} \cdot \dfrac{4}{1}$$
$$K = 3$$

2. Check for #1 $\quad \dfrac{1}{4}(3) - \dfrac{1}{2} = \dfrac{1}{4}$

$$\dfrac{3}{4} - \dfrac{1}{2} = \dfrac{1}{4}$$
$$\dfrac{1}{4} = \dfrac{1}{4}$$

3. $\dfrac{3}{5}X - \dfrac{1}{2} = 19\dfrac{1}{2}$

4. Check for #3

5. $\dfrac{1}{6}L + \dfrac{1}{3} = \dfrac{2}{3}$

6. Check for #5

7. $\dfrac{1}{2}Y + \dfrac{3}{4} = 7\dfrac{3}{4}$

8. Check for #7

9. $\dfrac{4}{5}M - \dfrac{1}{10} = \dfrac{3}{5}$

10. Check for #9

11. $\dfrac{1}{4}Z - \dfrac{2}{3} = 3\dfrac{1}{3}$

12. Check for #11

Use multiplication, addition, and subtraction as needed to solve each equation. Check your work by replacing the unknown with the solution.

1. $\dfrac{1}{2}P - \dfrac{1}{8} = \dfrac{1}{4}$

2. Check for #1

3. $\dfrac{1}{3}B - \dfrac{2}{7} = 1\dfrac{5}{7}$

4. Check for #3

5. $\dfrac{3}{8}N + \dfrac{1}{4} = \dfrac{5}{8}$

6. Check for #5

7. $\dfrac{4}{5}A + \dfrac{1}{9} = 4\dfrac{1}{9}$

8. Check for #7

9. $\dfrac{3}{4}R - \dfrac{1}{2} = \dfrac{1}{8}$

10. Check for #9

11. $\dfrac{1}{6}D - \dfrac{1}{6} = 8\dfrac{5}{6}$

12. Check for #11

Use multiplication, addition, and subtraction as needed to solve each equation. Check your work by replacing the unknown with the solution.

1. $\dfrac{5}{6}T - \dfrac{1}{3} = \dfrac{1}{3}$

2. Check for #1

3. $\dfrac{3}{4}F - \dfrac{3}{5} = 17\dfrac{2}{5}$

4. Check for #3

5. $\dfrac{2}{3}Q + \dfrac{1}{6} = \dfrac{1}{3}$

6. Check for #5

7. $\dfrac{3}{4}C + \dfrac{1}{3} = 15\dfrac{1}{3}$

8. Check for #7

9. $\dfrac{3}{5}S + \dfrac{7}{10} = \dfrac{9}{10}$

10. Check for #9

11. $\dfrac{1}{6}H - \dfrac{1}{3} = 1\dfrac{2}{3}$

12. Check for #11

Use multiplication, addition, and subtraction as needed to solve each equation. Check your work by replacing the unknown with the solution.

1. $\dfrac{3}{5}G + \dfrac{1}{2} = 9\dfrac{1}{2}$

2. Check for #1

3. $\dfrac{4}{7}U + \dfrac{2}{7} = \dfrac{6}{7}$

4. Check for #3

Write each fraction in hundredths. Then write it as a decimal and as a percent.

5. $\dfrac{2}{5} = \dfrac{}{100} = \underline{\quad\quad} = \underline{\quad\quad}\%$

6. $\dfrac{3}{4} = \dfrac{}{100} = \underline{\quad\quad} = \underline{\quad\quad}\%$

Use the Rule of Four to make denominators the same and then compare the fractions.

7. $\dfrac{5}{8} \bigcirc \dfrac{2}{3}$

8. $\dfrac{1}{5} \bigcirc \dfrac{2}{10}$

9. $\dfrac{3}{4} \bigcirc \dfrac{9}{16}$

QUICK REVIEW

In Roman numerals, L stands for 50, C stands for 100, D stands for 500, and M stands for 1,000. The Roman numerals C and M may be used up to three times in a row, but L and D may be used only once. C may be written before D or M to indicate subtraction. These symbols can be combined with the symbols reviewed in the last lesson to write any number up to 3,999. Study the chart.

40	XL	90	XC	300	CCC	700	DCC
50	L	100	C	350	CCCL	800	DCCC
60	LX	150	CL	400	CD	900	CM
70	LXX	200	CC	500	D	1,000	M
80	LXXX	250	CCL	600	DC	3,000	MMM

Write the Arabic numeral that corresponds to each Roman numeral. The first one has been done for you.

10. LXVIII __68__

11. CCXXX _____

12. DXXV _____

13. MDCX _____

Write the Roman numeral that corresponds to each Arabic numeral.

14. 49 _____

15. 352 _____

16. 583 _____

17. 2,555 _____

18. Devan dug a rectangular hole that was $3\frac{1}{2}$ feet long, $2\frac{1}{2}$ feet wide, and $1\frac{1}{2}$ feet deep. What was the volume of the dirt removed from the hole?

SYSTEMATIC REVIEW

Use multiplication, addition, and subtraction as needed to solve each equation. Check your work by replacing the unknown with the solution.

1. $\dfrac{1}{3}E + \dfrac{5}{8} = 6\dfrac{5}{8}$

2. Check for #1

3. $\dfrac{1}{8}V - \dfrac{1}{8} = \dfrac{3}{4}$

4. Check for #3

Write each fraction in hundredths. Then write it as a decimal and as a percent.

5. $\dfrac{1}{4} = \dfrac{}{100} = \rule{1cm}{0.4pt} = \rule{1cm}{0.4pt}\%$

6. $\dfrac{4}{5} = \dfrac{}{100} = \rule{1cm}{0.4pt} = \rule{1cm}{0.4pt}\%$

Multiply the numerators and denominators by the same number to make equivalent fractions.

7. $\dfrac{2}{3} = \rule{1cm}{0.4pt} = \rule{1cm}{0.4pt} = \dfrac{}{12}$

8. $\dfrac{5}{8} = \rule{1cm}{0.4pt} = \dfrac{15}{} = \rule{1cm}{0.4pt}$

Write the Arabic numeral that corresponds to each Roman numeral.

9. LXXXVI _____

10. CLII _____

11. MMMD _____

Write the Roman numeral that corresponds to each Arabic numeral.

12. 74 _____

13. 211 _____

14. 1,522 _____

Fill in the blanks to review measurement.

15. 1 yd = ____ ft

16. 1 qt = ____ pt

17. 1 gal = ____ qt

18. What number is equivalent to 5^2?

19. Solve for the unknown: $4X + 16 = 64$

20. What is the perimeter of a square that measures $5\frac{3}{8}$ inches on each side?

Use multiplication, addition, and subtraction as needed to solve each equation. Check your work by replacing the unknown with the solution.

1. $\frac{5}{6}J + \frac{2}{5} = 10\frac{2}{5}$

2. Check for #1

Write each fraction in hundredths. Then write it as a decimal and as a percent.

3. $\frac{3}{5} = \frac{}{100} = \underline{\hphantom{xxx}} = \underline{\hphantom{xxx}} \%$

4. $\frac{1}{2} = \frac{}{100} = \underline{\hphantom{xxx}} = \underline{\hphantom{xxx}} \%$

Follow the signs.

5. $2\frac{2}{15} \div 1\frac{1}{3} = \underline{\hphantom{xxx}}$

6. $2\frac{2}{5} \times 1\frac{1}{4} \times 1\frac{1}{7} = \underline{\hphantom{xxx}}$

7. $\begin{array}{r} 7\frac{1}{2} \\ -\ 3\frac{4}{5} \\ \hline \end{array}$

8. $\begin{array}{r} 3\frac{3}{8} \\ +\ 1\frac{4}{5} \\ \hline \end{array}$

Write the Arabic numeral that corresponds to each Roman numeral.

9. XCI _____

10. CDX _____

11. MCXXXV _____

Write the Roman numeral that corresponds to each Arabic numeral.

12. 58 _____

13. 412 _____

14. 2,260 _____

Fill in the blanks to review measurement.

15. 1 lb = ____ oz

16. 1 ft = ____ in

17. 1 mi = _____ ft

18. 1 ton = _____ lb

19. What is the approximate area of a circle with a radius of 35 miles?

20. What is the approximate circumference of a circle with a radius of 35 miles?

A fraction number line is a good way to review equivalent fractions. This number line may remind you of the rulers pictured in lesson 14 of the *Epsilon Instruction Manual*.

Write the missing names of the fractions in the boxes.

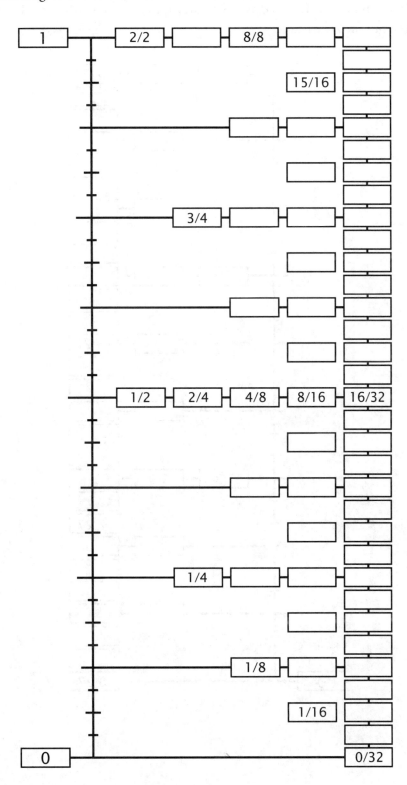

It is interesting to note that, no matter how many times you divide the space between two fractions, you can always divide the resulting space another time. This means there are an infinite number of fractions between 0 and 1. We won't ask you to write them all!

The fraction number line on the previous page was divided by multiples of two each time. Here is a fraction number line that is divided by multiplies of three. Most of the fraction names are left for you to put into the boxes.

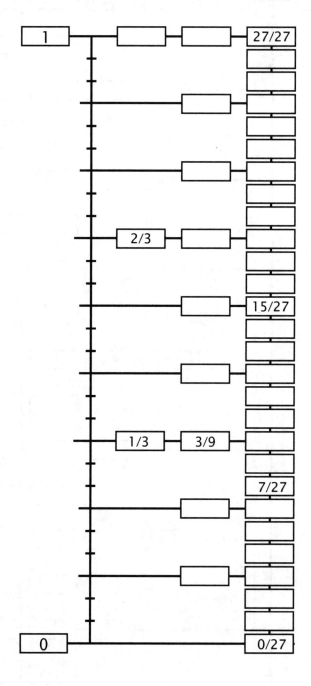

Find the area of the trapezoids. The first one has been done for you. Remember that the drawings are sketches and may not be drawn to scale.

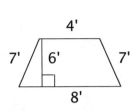

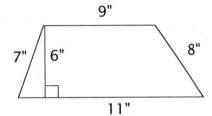

4 ft + 8 ft = 12 ft

12 ft ÷ 2 = 6 ft

6 ft × 6 ft = 36 sq ft

1. A = <u>36 sq ft</u>

2. A = _____

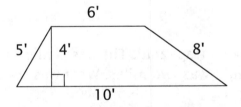

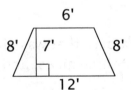

3. A = _____

4. A = _____

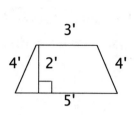

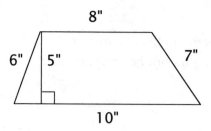

5. A = _____

6. A = _____

7. Austin cut shapes from colored paper to decorate the walls of his room. Each trapezoid had bases of seven and nine inches and a height of five inches. What was the area of each trapezoid?

8. Sam visited a park that was shaped like a trapezoid. The bases measured one mile and three miles, and the height was two miles. What was the area of the park?

Find the area of the trapezoids.

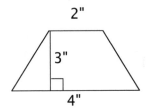

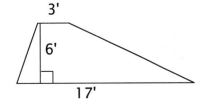

1. A = _____

2. A = _____

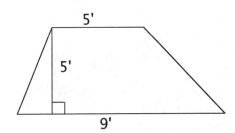

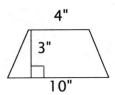

3. A = _____

4. A = _____

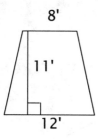

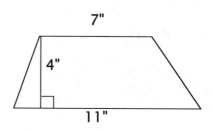

5. A =_____

6. A =_____

7. Pam's flower garden is shaped like a trapezoid. The bases are two and six feet long, and the height is six feet. The plants she plans to use need one square foot of space apiece. How many plants should Pam buy to fill the flower garden?

8. Richard designed a special little book shaped like a trapezoid. The top of the book is five inches wide, the bottom of the book is seven inches wide, and the height of the book is six inches. What is the area of the front cover of the book he designed?

Symbols and Tables

SYMBOLS

=	equals
≈	approximately equal
<	less than
>	greater than
%	percent
π	pi (22/7 or 3.14)
r^2	r squared or r · r
'	foot
"	inch
∟	right angle

PERIMETER

Any figure – add the lengths of all the sides

CIRCUMFERENCE

circle $C = 2\pi r$

AREA

rectangle, square, parallelogram
   $A = bh$ (base times height)

triangle $A = \dfrac{bh}{2}$

circle $A = \pi r^2$

trapezoid $A = \dfrac{b_1 + b_2}{2} \times h$

   (average of two bases times height)

EXPANDED NOTATION

$1{,}452.5 = 1 \times 1{,}000 + 4 \times 100 + 5 \times 50 + 2 \times 1 + 5 \times \frac{1}{10}$

MEASUREMENT

3 teaspoons (tsp) = 1 tablespoon (Tbsp)
2 pints (pt) = 1 quart (qt)
8 pints (pt) = 1 gallon (gal)
4 quarts (qt) = 1 gallon (gal)
12 inches (in) = 1 foot (ft)
3 feet (ft) = 1 yard (yd)
5,280 feet (ft) = 1 mile (mi)
16 ounces (oz) = 1 pound (lb)
2,000 pounds (lb) = 1 ton
60 seconds = 1 minute
60 minutes = 1 hour
7 days = 1 week
365 days = 1 year
52 weeks = 1 year
12 months = 1 year
100 years = 1 century
1 dozen = 12

DIVISIBILITY RULES

Number is divisible by:

2	if it ends in an even number
3	if digits add to a multiple of 3
5	if it ends in 5 or 0
9	if digits add to a multiple of 9
10	if it ends in 0

VOLUME

rectangular solid $V = Bh$
   (area of base times height)
cube $V = Bh$

Glossary

A–D

Arabic numerals - the numbers created by using the digits 0 through 9 in specific place values

area - the measure of the space covered by a plane shape, expressed in square units

average - a measure of center in a set of numbers, usually referring to the *mean*

base - a particular side or face of a geometric figure used to calculate area or volume

circumference - the distance around the outside of a circle

coefficient - a quantity placed before and multiplying the variable in an algebraic expression

composite number - a number with more than two factors

congruent - having exactly the same size and shape

cube - a solid with six congruent faces that meet at right angles

decimal (fraction) - a fraction written using a decimal point and place value

decimal point - a dot used to separate whole numbers and fractions; also used to separate dollars and cents

decompose - to separate a number into parts

denominator - the bottom number in a fraction, which shows the number of parts in the whole

dimension - a measurement in a particular direction (length, width, height, depth)

dividend - the number being divided

divisor - a number that is being divided into another

E–G

equation - a mathematical statement that uses an equal sign to show that two expressions have the same value

equivalent - having the same value

estimate - a close approximation of an actual value

even number - any number that can be evenly divided by two

expanded notation - a way of writing numbers by showing each digit multiplied by its place value

factor - (n) a whole number that multiplies with another to form a product; (v) to find the factors of a given product

factor tree - a diagram used to find the prime factors of a composite number

fraction - a number indicating part of a whole

greatest common factor (GCF) - the greatest number that will divide evenly into two or more numbers

H–P

height - the perpendicular distance from the base to the top of a figure

431

improper fraction - a fraction with a numerator greater than its denominator

inequality - a mathematical statement showing that two expressions have different values

least common multiple (LCM) - the least number that is a multiple of two or more other numbers

mixed number - a number written as a whole number and a fraction

multiplicative inverse - the number that, when multiplied by a given number, has a product of 1; reciprocal

numerator - the top number in a fraction, which shows the number of parts being considered

odd number - any number that cannot be divided evenly by two

percent - a ratio with 100 as the second part; shown with the symbol %

pi - the Greek letter π, which represents an irrational number with an approximate value of 22/7 or 3.14

place value - the position of a digit which indicates its assigned value

prime factorization - renaming a number as a product of two or more prime numbers

prime factors - all the factors of a number that are prime numbers

prime number - a number that has only two factors: one and itself

quadrilateral - a polygon with four sides

quotient - the result when numbers are divided

R–S

radius - the distance from the center of a circle to its edge; in a regular polygon, the distance from the center to any vertex; in a sphere, the distance from the center to any point on the surface; plural is *radii*

ratio - the relationship between two values; can be written in fractional form

reciprocal - the number that, when multiplied by a given number, has a product of 1; multiplicative inverse

rectangle - a quadrilateral with two pairs of opposite parallel sides and four right angles

rectangular solid - a three-dimensional shape with six rectangular faces

regrouping - composing or decomposing groups of ten when adding or subtracting

repeated division - a method for finding the prime factors of a number

right angle - an angle measuring 90 degrees

Roman numerals - a system used by the ancient Romans in which letters represent numbers

rounding - replacing a number with another that has approximately the same value but is easier to use

"Rule of Four" - a Math-U-See method for finding the common denominator of two fractions

"same-difference theorem" - a Math-U-See method for subtraction that adds the same value to minuend and subtrahend to avoid regrouping

simplify - to rewrite an expression as simply as possible; a simplified fraction will have a numerator and denominator with a single common factor of one

square - a quadrilateral in which the four sides are perpendicular and congruent

T–Z

trapezoid - a four-sided polygon with a set of parallel sides

unit - the place in a place-value system representing numbers less than the base

unit fraction - a fraction with a numerator of one

unknown - a specific quantity that has not yet been determined, usually represented by a letter

volume - the number of cubic units that can be contained in a solid